Anonymous

Haverhill, Massachusetts

An industrial and commercial center

Anonymous

Haverhill, Massachusetts
An industrial and commercial center

ISBN/EAN: 9783337328368

Printed in Europe, USA, Canada, Australia, Japan

Cover: Foto ©berggeist007 / pixelio.de

More available books at **www.hansebooks.com**

HAVERHILL

MASSACHUSETTS

AN INDUSTRIAL

AND

COMMERCIAL CENTER

PUBLISHED BY THE BOARD OF TRADE

HAVERHILL MASSACHUSETTS
1889
CHASE BROTHERS

CONTENTS.

PREFACE.

This book has been prepared by the Committee on Statistics and Information of the Board of Trade of Haverhill in obedience to instructions from the Board.

They have meant to indulge little in reminiscence, but it has been their aim to present such a picture of the present Haverhill as not only to remind its own citizens of what easily slips the memory of the most loyal, but also to give the stranger an adequate conception of its claims to rank among the chief industrial cities of the country; of its origin, its progress, and its yet undeveloped possibilities; of its success, its natural beauty, its hospitality, its energy, its organic life.

It has been their endeavor to state only what are conceded to be facts, in the belief that the recital of the facts alone invites to Haverhill the capitalist, the manufacturer, the tradesman, and the artisan, the man of means looking for a reasonable investment, the man of family in quest of a home, the man of leisure in search of a refined society, the parent solicitous for the welfare of his children.

To these, this imperfect sketch of Haverhill and its industries, imperfect because done by the busy residents of a busy city, is presented. If it shall do no more than create a closer acquaintanceship between these and the citizens of Haverhill, it will have served a worthy end.

HAVERHILL

BOARD OF TRADE.

Pursuant to a call for a meeting of business men and citizens of Haverhill to consider the propriety of the formation of a Board of Trade, about fifty gentlemen met in the office of George A. Hall, Esq., Academy of Music, March 30, 1888, and organized by the choice of H. E. Bartlett, chairman, and E. G. Frothingham, secretary. A committee was appointed to nominate a list of officers for permanent organization and to prepare a constitution and by-laws, which committee met at an adjourned meeting at No. 40 Daggett's Building, April 2, and voted to recommend for adoption a constitution and by-laws, and nominated a list of officers, all of which action was accepted and confirmed at the first regular meeting of the Board, held at the Police Court Room, April 11, 1888.

PRESIDENT, THOMAS SANDERS.

VICE PRESIDENTS.

Levi Taylor.
James H. Carleton,
George A. Kimball,
J. B. Swett,
John B. Nichols,
Daniel Fitts,
John L. Hobson,
A. W. Downing,
John E. Gale,
Daniel Goodrich,
Gyles Merrill,
Warren Hoyt,
A. A. Hill,
George H. Carleton,
Dudley Porter,
B. F. Brickett,
B. F. Leighton,
James O'Doherty,
L. V. Spaulding,
Charles Butters,
Alpheus Currier,
Charles W. Chase,
E. B. Bishop,
A. P. Jaques,
Charles H. Goodwin,
C. N. Kelly,
C. H. Weeks,
Charles Shapleigh,

Charles S. Kendrick,
George Thayer,
F. E. Watson,
George O. Willey,
George O. Hoyt,
H. B. Goodrich,
J. H. Sayward,
George C. How,
E. O. Bullock,
E. H. Howes,
J. C. Hardy,
F. G. Richards,
W. H. Smiley,
George A. Greene,
S. P. Gardner,
C. P. Messer,
J. J. Winn,
I. B. Hosford,
Alfred Kimball,
Ira O. Sawyer,
Henry Belanger,
John A. Gale,
D. D. Chase,
Ira A. Abbott,
A. M. Allen,
Alonzo Way,
Warren Emerson,
J. A. Huntington,

Irah E. Chase,
W. H. Moody,
Algernon P. Nichols,
J. H. Sheldon,
Charles Smiley,
Albert LeBosquet,

Charles LeBosquet,
U. A. Killam,
L. C. Wadleigh,
W. E. Blunt,
W. R. Whittier,
A. M. Tilton.

C. H. Fellows,

DIRECTORS.

William A. Brooks,
James H. Winchell,
George A. Hall,
B. B. Jones,
J. G. S. Little,
C. W. Morse,
M. W. Hanscom,
D. F. Sprague,

Charles W. Arnold,
Martin Taylor,
Charles N. Hoyt,
Aug. Bourneuf,
Woodbury Noyes,
George L. Emerson,
T. S. Ruddock,
D. T. Kennedy,

F. C. Wilson.

TREASURER, HORACE E. BARTLETT.

SECRETARY, E. G. FROTHINGHAM.

STANDING COMMITTEES.

FINANCE AND ROOMS.

U. A. Killam, Chairman.

D. F. Sprague,   C. W. Arnold.

RAILROADS AND TRANSPORTATION.

George H. Carleton, Chairman.

B. B. Jones, Secretary.

J. H. Winchell,   Alfred Kimball,

Martin Taylor.

MUNICIPAL AFFAIRS.

E. B. Bishop, Chairman.

Thomas E. Burnham, Secretary.

Woodbury Noyes,   James D. White,
Charles N. Kelly.

MANUFACTURING AND MERCANTILE AFFAIRS.

W. A. Brooks, Chairman.

C. W. Morse, Secretary.

Thomas S. Ruddock,   F. C. Came,
George C. How,   Ira O. Sawyer,
F. G. Richards.

STATISTICS AND INFORMATION.

Jones Frankle, Chairman.

W. E. How, Secretary.

A. A. Hill,   M. D. Clarke,
J. J. Winn.

MEMBERS.

George H. Appleton,   N. K. Johnson,
A. H. Adams,   J. E. Kimball,
Walter Ayer,   L. Kimball & Son,
Thomas H. Bailey,   B. M. Kimball & Son,
S. C. Bassett,   N. S. Kimball,
William Bray,   Warren Kimball,
C. I. Bickum,   J. E. Lord,
A. C. Barrows,   Thomas Lahey,
W. T. Barstow,   W. B. Lamprey,
W. F. Blake,   B. T. Longfellow,
Bennett & Co.,   J. A. Lynch,

B. F. Barnes,
Chester Bryant,
Hiram Bond,
R. G. W. Butters,
B. A. Ball,
George Brooks,
M. Bonin,
J. C. Bates,
H. E. Chase,
A. W. Cram,
C. Haven Coffin,
F. A. Cheney,
E. Charlesworth,
A. Wash. Chase,
H. W. Chase,
C. W. Chandler,
F. H. Cate,
F. C. Came,
John A. Colby,
C. H. Cushman,
Maurice D. Clarke,
Clark & Dow,
L. H. Chick,
H. M. Clay,
George H. Cleveland,
Thomas F. Carroll,
George B. Case,
Charles T. Chase,
R. S. Chase,
W. D. Collins,
Chase & Meader,

George W. Ladd,
William Lyall,
George V. Ladd,
I. L. Mitchell,
W. S. Merryman,
F. J. Mitchell,
C. C. Morse & Son,
E. A. Mitchell,
L. E. Martin,
J. K. Moody,
Eben Mitchell,
H. F. Morse,
J. J. Marsh,
William Nason,
Byron Noyes,
C. C. Osgood,
A. A. Ordway,
J. H. Osgood,
Charles T. Paul,
E. H. Pinkham,
G. W. Pettingill,
Edwin Poor & Co.,
Nicholas Powers,
J. W. Proctor,
A. D. Patch,
H. I. Pinkham,
John Pilling,
F. A. Pierce,
W. H. Page,
H. L. Perkins,
Harvey Ray,

J. M. Davis,
B. C. Davis,
S. A. Dow.
John Duncan, Jr.,
H. L. Dole,
James Dewhirst,
Robert Driscoll,
C. Willis Damon,
Moses H. Dow,
W. F. Endicott,
W. F. Evans,
I. H. Eaton,
Charles Edwards,
Luther Emerson,
E. H. Emerson,
C. B. Emerson,
Matthew French,
Floyd & Peabody,
E. A. Fitts,
A. E. Fernald,
W. M. Fellows,
Jones Frankle,
C. K. Fox,
C. H. Gleason,
J. W. Goodwin,
W. S. Goodell,
J. N. B. Green,
J. A. Gage,
H. H. Gilman,
M. S. Holmes,
Moses How,

Frank H. Russ,
Russell & Co.,
J. W. Russ,
F. L. Ricker,
C. N. Rhodes,
George W. Russ,
Joseph Ridgeway,
Perley A. Stone,
W. W. Spaulding,
William Sawyer,
Charles H. Smith,
E. L. Shannon,
A. H. Saltmarsh,
D. Sherwood,
J. M. Stover,
P. C. Swett,
W. K. Stratton,
J. F. Smith,
J. B. Simas,
M. L. Stover,
F. E. Tucker,
H. C. Tanner,
Thomas J. Taylor,
J. R. Thing,
W. B. Thom,
C. R. Thom,
D. B. Tenney,
J. M. Taylor,
E. G. Tilton,
George H. Tilton,
W. H. Underhill,

Daniel Hooke,
J. W. Hayes,
James A. Hale,
George H. Hill,
C. D. Hunking,
J. M. Haseltine,
George W. Hanson,
W. C. Hunkins,
A. J. Hodgdon,
E. C. Holman,
Hoyt & Taylor,

Varney & Hayes.
J. H. Varney.
George W. Wentworth.
J. F. West.
James D. White.
C. T. Weaver,
D. R. Webster,
J. O. Wardwell,
L. C. Wadleigh, Jr..
L. J. Young.
A. B. Jaques.

Historic Haverhill.

Haverhill is the child of destiny. An inland village on the Merrimack, wanting the steep waterfalls of the upper river and the harbor of the lower, a stranger to the capricious and unexpected leaps in growth of other manufacturing centers, it has pursued its way in steadfastness, until the settlement of the Puritans became a village, the village grew into a town, and the town unfolded into the Haverhill of to-day, — an aggressive, substantial, energetic, thriving city, conservative in its clasp of the past, radical in its reach for the future. Little could Goodman Ward, rowing up the river that summer day two hundred and fifty years ago, imagine that the log hut he was to build held the germ of to-day's city, with its factories and blocks, its steam and horse railways, its electric lights, its telegraphs and telephones, its fire department and water-works, the very invention of most of which was not yet dreamed of.

Haverhill, or Pentucket, as the Indians called the
spot, was begun in 1640 by a fragment of the emigrants
who replanted in Massachusetts the English Essex,
bringing with them the sturdiness, integrity, and love
of freedom indigenous to their birthplace, and recall-
ing their old homes in the names they gave the new.
Thus, in honor of the native town of their leader and
first minister, the English Haverhill was commemorated
by the founders of the new. Honorable in their earliest dealings with the aborigines, they bought of the
Indians the lands they sought to occupy, the original
deed being still preserved in the city's archives, an
evidence of good faith on the part of some, at least, of
the foreign trespassers upon these shores.

Honest dealings with the owners of the soil did not,
however, protect the villagers of the earlier days from
the oft-repeated attacks of hostile Indians. Haverhill
occupied a peculiar position in this regard, lying on the
outermost edge of the settlements and being thus more
directly exposed to the fury or vindictiveness of the
hostile bands that swept down the valley of the Merrimack or across the country. For nearly a century
Haverhill suffered from the repeated forays of the savages, being for the first fifty years in daily expectation
of an attack. At length, however, other towns grew
upon its northern borders and stood between it and its
savage foes. There still remain, in various parts of the
city, as the instinct of safety suggested their erection,
garrison houses, so called, whither the adjacent settlers
were in the habit of betaking themselves upon the first
suggestion of hostile approach. Of brick, to guard

against being set on fire, of good size, to afford safe
retreat for the endangered settlers, with convenient
loop-holes, they afford substantial and undeniable suggestion of the danger and the heroism of the lives
our forefathers led in the wilderness.

Memorable in Haverhill, and celebrated then and
since far beyond the town's horizon, were the adven-
tures attending the capture and escape of Hannah
Duston. On the fifteenth of March, 1697, a body of
Indians made an unexpected descent upon the town and
came to the house of Thomas Duston, who was living
in one of the outlying settlements. "This man was
abroad at his usual labour. Upon the first alarm, he
flew to the house, with the hope of hurrying to a place
of safety his family, consisting of his wife, who had
been confined a week only in child-bed, her nurse, a
widow from the neighborhood, and eight children.
Seven of his children he ordered to flee with the
utmost expedition in the course opposite to that in
which the danger was approaching, and went himself
to assist his wife. Before she could leave her bed, the
savages were upon them. Her husband, despairing of
rendering her any service, flew to the door, mounted
his horse, and determined to snatch up the child with
which he was unable to part when he should overtake
the little flock. When he came up to them, about two
hundred yards from his house, he was unable to make
a choice or to leave any one of the number. He
therefore determined to take his lot with them, and to
defend them from their murderers or die by their side.
A body of the Indians pursued and came up with him,

and from near distances fired at him and his little company. He returned the fire and retreated, alternately.
For more than a mile he kept so resolute a face to his
enemy, retiring in the rear of his charge, returned the
fire of his enemies so often and with so good success,
and sheltered so effectually his terrified companions,
that he finally lodged them all safe from the pursuing
butchers in a distant house. When it is remembered
how numerous his assailants were, how bold, when an
over-match for their enemies, how active, and what
excellent marksmen, a devout mind will consider
the hand of Providence as unusually visible in the
preservation of this family.

"Another part of the Indians entered the house
immediately after Mr. Duston had quitted it, and found
Mrs. Duston and her nurse, who was attempting to fly
with the infant in her arms. Mrs. Duston they ordered
to rise instantly, and, before she could completely dress
herself, obliged her and her companion to quit the
house, after they had plundered it and set it on fire.
In company with several other captives, they began
their march into the wilderness, she feeble, sick, terrified beyond measure, partially clad, one of her feet
bare, and the season utterly unfit for comfortable traveling. The air was chilly and keen, and the earth
covered, alternately, with snow and deep mud. Her
conductors were unfeeling, insolent, and revengeful.
Murder was their glory and torture their sport. Her
infant was in her nurse's arms, and infants were the
customary victims of savage barbarity. The company
proceeded but a short distance, when an Indian, think-

ing it an incumbrance, took the child out of the nurse's
arms and dashed its head against a tree. What were
then the feelings of the mother?

"Such of the other captives as began to be weary
and to lag, the Indians tomahawked. The slaughter
was not an act of revenge or cruelty. It was a mere
convenience; an effort so familiar as not even to ex-
cite an emotion. Feeble as Mrs. Duston was, both she
and her nurse sustained, without yielding, the fatigue
of the journey. Their intense distress for the death of
the child and of their companions, anxiety for those
whom they had left behind, and unceasing terror for
themselves raised these unhappy women to such a degree of vigour, that, notwithstanding their fatigue,
their exposure to cold, their sufferance of hunger, and
their sleeping on damp ground under an inclement sky,
they finished an expedition of about one hundred and
fifty miles, without losing their spirits or injuring their
health. The weekwam to which they were conducted
and which belonged to the savage who had claimed
them as his property was inhabited by twelve persons.
In the month of April this family set out with their
captives for an Indian settlement still more remote, and
informed them, that, when they arrived at the settlement, they must be stripped, scourged, and run the
gauntlet, naked between two files of Indians, containing the whole number found in the settlement; for
such, they declared, was the standing custom of their
nation. This information, you will believe, made a
deep impression on the minds of the captive women,
and led them, irresistibly, to devise all the possible

means of escape. On the thirty-first of the same
month, very early in the morning, Mrs. Duston, while
the Indians were asleep, having awaked her nurse and
a fellow-prisoner (a youth taken some time before
from Worcester), dispatched, with the assistance of
her companions, ten of the twelve Indians. The other
two escaped. With the scalps of these savages they
returned through the wilderness; and, having arrived
safely at Haverhill, and afterwards at Boston, received
a handsome reward for their intrepid conduct from the
legislature." A monument on the common, close to
the site of the old meeting-house, commemorates the
event.

Another day whose mournful events have been preserved in both history and tradition was the twenty-
ninth of August, 1708, when Haverhill was attacked
by a band of French Indians, recruited in Canada.

"At break of day they passed the frontier garrisons
undiscovered, and were first seen near the pound,
marching two and two, by John Keezar, who was re-
turning from Amesbury. He immediately ran into the
village and alarmed the inhabitants, who seem to
have slept totally unguarded, by firing his gun near the
meeting-house. The enemy soon appeared, making
the air ring with terrific yells, with a sort of whistle,
which, says tradition, could be heard as far as a horn,
and clothed in all the terrors of a savage war-dress.
They scattered in every direction over the village, so
that they might accomplish their bloody work with
more despatch. The first person they saw was a Mrs.
Smith, whom they shot as she was flying from her

house to a garrison. The foremost party attacked the house of Rev. Benjamin Rolfe (the second minister of the place), which was then garrisoned with three soldiers; and he and a part of his beloved family were suddenly awakened from their slumbers only to hear the horrid knell for their departure. Mr. Rolfe instantly leaped from his bed, placed himself against the door, which they were endeavoring to beat in, and called on the soldiers for assistance; but these craven-hearted men refused to give it, for they were palsied with fear and walked to and fro through the chambers, crying and swinging their arms. Had they displayed but half the ordinary courage of men, no doubt they would have successfully defended the house. But, instead of that, they did not fire a gun or even lift a finger towards its defence. The enemy, finding their entrance strenuously opposed, fired two balls through the door, one of which took effect and wounded Mr. Rolfe in the elbow. They then pressed against it with their united strength, and Mr. Rolfe, finding it impossible to resist them any longer, fled precipitately through the house and out at the back door. The Indians followed, overtook him at the well and dispatch him with their tomahawks. They then searched every part of the house for plunder, and also for other victims, on whom they might inflict their savage cruelty. They soon found Mrs. Rolfe and her youngest child, Mehitabel; and, while one of them sunk his hatchet deep in her head, another took the infant from her dying grasp and dashed its head against a stone near the door. Two of Mr. Rolfe's children, about

six and eight years of age, were providentially saved
by the sagacity and courage of Hagar, a negro slave,
who was an inmate of the family. Upon the first
alarm, she leaped from her bed, carried them into the
cellar, covered them with two tubs, and then concealed herself. The enemy entered the cellar and
plundered it of everything valuable. They repeatedly
passed the tubs that covered the two children, and
even trod on the foot of one, without discovering
them. They drank milk from the pans, then dashed
them on the cellar bottom, and took meat from the
barrel behind which Hagar was concealed." The
three soldiers obtained nothing by their cowardice, as
they plead for mercy in vain.

While these, the central figures of the tragic day,
were thus engaged, the remainder of the attacking
party had been finding other victims, among whom
were women and children, the captain of the town
militia, and the first selectman. Between thirty and
forty were killed or taken prisoners. Several dwellings
were burned, and an attempt made to destroy the
meeting-house, but this was frustrated by the coolness
of one man who raised the cry that help was at hand.
The Indians were thus panic-stricken before they had
done what mischief they might. By this time a force
of soldiers and of the townspeople had been collected
and pursued the enemy, who had left the town precipitately. They came up with them two miles away and
attacked them, although inferior in numbers; and, after
a skirmish of about an hour, the Indians fled, leaving
nine dead and carrying off several wounded. Many of

the prisoners and most of the plunder were recovered.
Some of the prisoners were barbarously slain to prevent their escape. The inhabitants were left to the
sorrowful office of burying their dead. The day was
somewhat advanced when the battle was over, and, it
being extremely warm, the interment was necessarily
hurried. Coffins could not be made for all, and a
large pit was dug in the burying-ground, in which several were laid. Some of those who fell in the last engagement were, it is supposed, buried on the spot.
This was the last, as it was the most formidable, attack of any importance made by the Indians upon the
town. There were marauders now and then, and occasional alarms, but they grew less and less as time
wore on.

There was little of the sensational or startling, beyond the constant menace of the Indians, in the town's
early days. The few first settlers multiplied by natural increase and by additions from without. They
robbed the primal wilderness of its wooded intervals
and turned them into corn-fields. They fed their families on the fish — salmon, shad, and alewives — with
which the Merrimack (river of sturgeons, as some
have translated it) ran thick. Though the men from
Newbury who broke ground in Haverhill came up the
river in 1640, it was not until 1642 that they acquired a
title to the land they tilled by the purchase from the
Indians already referred to. In 1643 the first town
meeting was held, and then was the first reference to
the disposition of the territory thus acquired, which
gave in later years no end of trouble, and was a very

important and practical matter in the affairs of the town.

"The theory of ownership and distribution of lands was apparently the following: The townsmen of that time had, by foresight, energy, and influence, obtained leave of the General Court to begin a plantation in a most desirable location. They had fairly purchased of the Indians a very large tract of territory. They held it legally and equitably, subject to the demands of the general government for the common weal, and the adjustment of bounds between them and their neighbors by competent authority. It was their property. They were the proprietors. They could divide it at such times and in such proportions as they saw fit. Such parts of it as were allotted to any particular one of them, he and his heirs and assigns would thereafter own in severalty. In other words, the persons then and there settled were 'ye inhabitants of Pentuckett,' to whom the Indians had sold. They had not bought for the benefit of all the persons who might flock to Pentuckett to profit by the advantageous grant they had obtained. If they chose, however, they could admit any person to their association and a participation in its privileges. And it must be said that the logic of the early settlers seems to have substantially prevailed. There came a time when their heirs and assigns assumed to be owners of all the lands remaining undivided, and, although fiercely opposed, maintained their claim with ultimate success. They held 'proprietors'' meetings, had their clerk and moderator, kept records, made grants, carried on successful litiga-

tion, and had their own way. Then the organization
quietly died out."

As time wore on and the settlement began to bear
less the look of a "clearing" and more that of a village, a variety of trades and manufactures sprang up
and in time assumed more or less prominence. One
of the earliest to be established and one of the last to
be given up was that of tanning, but there is now no
leather made in Haverhill, although the vats of the
tanner stood open over two hundred years. Other industries, now lapsed into desuetude, were the manufacture of potash, of salt, of saltpeter, and of duck
cloth, brewing, and distilling. Ship-building, begun
one hundred and fifty years ago, was also carried on
with vigor and to an extent much larger than might be
supposed, reaching its period of greatest prosperity at
the beginning of this century. At that time there were
three ship-yards in the central village and another at
East Haverhill. The vessels were ships, brigs, sloops,
schooners, and there have have been three launched in
a day at the village. There was need of vessels. At
that time Haverhill was carrying on an extensive commerce, along the coast, to the West Indies, and to
England, ships sailing from Haverhill to London direct. The town exported corn, grain, beef, fish, lumber, pearl-ashes, linseed oil, etc., bringing home sugar
and molasses from the West Indies and goods of all
kinds from the mother country. The vessels, if not
too large, came up the river and discharged at Haverhill; otherwise they were unloaded at Newburyport,
where their cargoes were transferred to scows and thus

brought up stream. After a while the carrying trade
fell off and ship-building languished, coming virtually
to an end in 1810. Since then, in 1875, two vessels
have been launched at Haverhill, but no others have
been built here, and there is no prospect of any farther
employment for the shipwright's adze or the calker's
hammer. The first distillery was built when the town
was nearly a hundred years old, and it was about a
hundred years later, when the last of the several that
had been in active operation was bought by a prominent advocate of total abstinence and the fires put out
the same night.

The manufacture of hats has been and is extensively carried on in Haverhill. Begun at least a quarter of
a century before the Revolution, it has been maintained ever since. The shops are now reduced in
number, though the output is not lessened, to two or
three large establishments, where hats are made only
of wool and by the factory system. In the early part
of the century, however, when the business had got
well under way, there were many shops, scattered in
various parts of the town. Hats were then made of
the fur of the beaver, raccoon, and muskrat; of cotton,
with pasteboard bodies; of silk and "napped" fur as
well as of wool. In connection with the manufacture of
these goods, it is worth while to recall the primitive
manner in which they were got to market. They were
carried on horseback for many years, and, later, when
wheels were heard of in the town, were transported by
this means, suspended in boxes from the axles. As
late as 1804 there were but two horse-carts in town.

The most important and valuable of Haverhill's industries is, as all the world knows, the making of
shoes, which had its origin and growth here without
any set purpose, but by the accident of fate or by a
species of natural selection. Cities have risen from
the sand because of their proximity to abundant water-
power; the purity of water, the proximity of fuel,
the neighborhood of the sea, have determined the location of enterprises; this thing or that is manufactured where material is plenty, labor easily obtainable,
or freights cheap: but Haverhill has become the manufacturer of an immense number of shoes, at times the
largest manufacturer of the world, without peculiar
cause.   Like Topsy, it "grew so." It is on record
that the shoemaker met with no very warm reception
upon his first appearance in Pentucket and that those
of the craft who applied for citizenship were at times
refused.   But, as has been pointed out, it is probable
that it was not the shoemaker as such who was refused, but the class of which, unfortunately, the early
shoemaker was a type, — a wanderer from place to
place and with a wanderer's tastes and habits. The
cobbler was, nevertheless, an evident necessity, and
cobblers and shoemakers became, in the natural course
of events, citizens and residents of Haverhill. There
was nothing, however, in this result that suggested the
promise or potency of the prodigious development of
later days.

From the earliest times until about the beginning
of this century, shoemaking in Haverhill was confined
almost entirely to supplying the wants of the com-

munity itself. Shoes were not made up in quantities
and kept on hand for sale, like most kinds of
goods at the present day; much less were they manufactured for foreign consumption. The time is almost
within the memory of persons now living, when it was
the common custom, outside of the villages, for shoemakers to "whip the stump," i. e., go from house to
house, stopping at each long enough to make up a
year's supply for the family. Farmers usually kept a
supply of leather on hand for family use, and in many
cases they were their own cobblers. A few years ago
a very rich farmer died at a great age in another town
of the state who had never worn shoes not of his own
making. A farmer was sometimes, being perhaps
fonder of tools or handier with them, the shoemaker for
the whole neighborhood, and worked at making or
mending shoes on rainy days and during the winter
season.

In villages, the "village cobbler," or shoemaker,
gradually came to keep a little stock of leather on
hand, and to exchange shoes with the farmers, tanners,
traders, and others, for produce, leather, foreign goods,
etc. There are said to have been but two shoemakers
in Haverhill as late as 1794. In course of time, the
storekeepers, then carrying on a very large commerce
with a wide region round about, began to keep a few
shoes on hand for sale. This was a natural outgrowth
of the barter system of trade, then the chief method
of dealing. The owners of the great "country stores"
bartered with the shoemakers for their shoes, bartered
the shoes with the back country farmers for produce,

and then bartered the produce for English and West
India goods. So, in 1795, it came about that one of
the merchants of the place advertised, that he had
" several thousand" fresh and dry hides which he
would exchange for shoes, giving credit for the hides
until the shoes could be made out of them. And, in
course of time, the merchants, seeing the possibility of
gain, became themselves the makers of shoes as well
as the sellers. The country market soon proved too
limited, nor was there demand enough in Boston and
the lesser places on the coast, and, so, during the war
of 1812, one of the more enterprising manufacturers
sent a wagon-load of shoes to Philadelphia, from
which he is said to have obtained a handsome profit.
Later, goods were sent even farther south. And so
Haverhill fell into the way of making shoes, and a
good many of them, which demanded and obtained a
wide and distant market. The two-horse " baggage-
wagon," of the early " freighter" Slocomb, making
regular trips between Haverhill and Boston since 1818,
failed to supply the demands of an increasing traffic;
and he was obliged to increase his facilities until in
1836 he employed forty horses and eight oxen, and his
large covered wagons were said, with perhaps a trifle
of imagination, to have almost literally lined the thirty
miles of road. The main highway in many of the
towns intervening between Boston and Haverhill still
bears the name of Haverhill Street, unconsciously pre-
serving the traditions of the days when the drivers of
the shoe teams were the most frequent travelers and
roads pointed one way to Boston and the other to Ha-

verhill. In 1837 there were forty-two shoe manufacturers and fourteen tanners and leather dealers in town, but the financial panic of that year dealt a hard blow to the shoe industry, from which it did not recover until the discovery of California gold lent a new impetus to trade. In 1860 the number of shoe factories had increased to one hundred, and from that time on the growth of the town's chief interest has been reasonably uniform and steady, outside of the inevitable misfortunes entailed by the war of 1861. One of the oddest fashions of the earlier manufacture was the disregard of method in packing, shoes being packed and shipped for some years without any attention to the sizes or the number in a case.

Haverhill was so related geographically to towns near and distant, being in its early days, when Lawrence was not dreamed of, the only inland town of account upon the river from Newburyport to Lowell, and affording, at first by a well-known ferry (by which Washington crossed in his journey through Essex after the Revolution) and later by a famous bridge, convenient passage across the Merrimack, that all the tide of travel from " above " poured through it and into it, and its " general stores " were remarkable for their size, and the multifarious nature of their contents. Several lines of stage-coaches ran to Boston, while others made regular trips to Salem, Lowell, Newburyport, Exeter, Dover, and Concord, N. H. It was in those days, too, that the inns and taverns of the towns at which the coaches stopped earned a just prominence and reputation, the Eagle House of Haverhill being a typical

example. The same house still stands, devoted to the
same purposes, though the changing times have robbed
it of its former prestige. Afterwards, when other
towns had grown and other bridges had been built,
Haverhill yet retained its prominence as a trading
center, since the growth of the shoe manufacturing
industry made it the focus to which converged the lines
of travel from many points of the compass and from
great distances. The same influences made it at once
the market for the produce of the farms, the point
where their finished shoes found sale, and the empo-
rium where diverse needs could be supplied. Partly
from the force of habits once formed, partly on account
of the relations between shoemaking and the inhabitants of the country towns, and partly from the abundant opportunities its well-filled stores afford to all sorts
of seekers after all sorts of wares, Haverhill still retains
its position as the center of a circle whither streams of
trade tend like its radii. The times, under the influence of railway communication, have greatly changed
since the main street of the village used to be so
crowded with teams as to be almost impassable, the
owners having come in to deal at " the store," but, in
spite of railway and steamboat, express and postal service, the same tendency holds, and for miles back into
the country, in Essex County and in lower New
Hampshire, the dweller on farm or in village turns his
steps to Haverhill when in need of whatever his farm
or village fails to supply. While, therefore, for such
reasons Haverhill invites to itself these customers, the
fact of their coming reacts on the city itself, and neces-

sity, if nothing else, compels its merchants, if they would retain this enormous trade, to the possession of spacious and well-lit stores, enough and courteous clerks, an abundant assortment of wares at reasonable prices. These Haverhill has; and, therefore, it is not alone one of the largest manufacturers of shoes in the world, but the source and center of a vast and increasing domestic commerce, to the advantage both of buyer and seller and with the result of vastly increasing the diameter and circumference of the actual Haverhill.

Haverhill has never lacked for patriotic spirit when the occasion required. The town records bear witness to the loyalty to the cause, the willingness to spend, the readiness to do, that apparently came by just inheritance from the Indian-fighting forefathers. In all the proceedings of the colonies just precedent to the great struggle with the mother country Haverhill had its part. When the oppressive measures of taxation were ordered by the King, Haverhill held town meetings to deal with the matter; when the unjust proceedings were persisted in by Great Britian, Haverhill joined with other towns of spirit in "boycotting" foreign goods; and, when the Continental Congress was weighing the question of finally dissolving allegiance to the mother country, the men of Haverhill, like those of all other New England towns with rare exceptions, pledged themselves " with their lives and fortunes to support them in the measure." The news of the battle of Lexington reached Haverhill at noon of the day it was fought, and before night one hundred and five Haverhill men (almost one-half of the entire militia force of the town) were "gone to yᵉ army." In the battle of Bunker Hill fought seventy-four men from Haverhill, about one in twenty of the entire command, of whom two were killed. And the same spirit of devotion to the cause was displayed all through the seven years' war. "There was no evidence of grumbling or despondency," remarks a recent writer, "and the demands were very great; scarcely was one quota filled, when another was called for. There were so many emergencies that life must have seemed full of them and to contain nothing else." In one year the expenses of the town for soldiers were over fifty thousand dollars. Every soldier required by the constant drafts was furnished up to the close of the war with the exception of a single man.

The war of 1812 afforded renewed opportunities for the exhibition of the same patriotic spirit. Though many of the citizens condemned this second war with England as uncalled for and ill advised, and, though towns all about it had passed and were passing resolutions of censure and disapproval, yet no sooner had a call been made for soldiers, than the town met at a short twenty-four hours' notice and generously voted, in substance, that no man's poverty should bar his patriotism. A large number of Haverhill men enlisted. Nevertheless, the news of peace was very grateful; and the cessation of hostilities was celebrated by a day of general rejoicing, with the ringing of bells, firing of cannon, illumination of houses, and religious services.

Another consecration of money and of life to the service of the country was made during the late civil war. The scenes that before many months of the struggle had passed became so familiar in all the northern towns were early enacted in Haverhill. The youth volunteering for enlistment, the muster on the village green, the escort of admiring friends and neighbors, the bitter leave-taking at last. — Haverhill was among the first to witness these. On the twenty-fourth of the January previous to the war the local militia company had held a meeting and its members had pledged each other to be in readiness for immediate departure should the occasion arise, and so, on the day when the attack was made on the Massachusetts Sixth in Baltimore, they started for Washington on receipt of the news. Only three days later a soldiers' relief society was formed, which did much work and immeasurable good in the succeeding four years. Haverhill sent to the war about thirteen hundred men, eighty-five more than were claimed of it. Of these, seventy-three were mustered out as commissioned officers, of whom six were field officers. — three colonels, one lieutenant-colonel, and two majors. The town raised and expended over a hundred thousand dollars for the support of the war, exclusive of state aid, and spent an equal sum for the latter purpose, which was afterwards refunded by the state. Even in the closing months of the struggle the town authorized continued enlistments to anticipate a possible call by the President. During the war excitement ran high in Haverhill, and there were some turbulent scenes, during which the sympathizers with the South were rather roughly handled, one being ridden on a rail and covered with tar and feathers.

SOLDIERS' MONUMENT.

The town testified its appreciation of its citizens
who fell in the country's service by erecting a beautiful monument in their honor in one of the public
squares, to which it has given a name. It is twenty-
six feet in height, with a base, a plinth with buttresses
surmounted by inverted cannon, and a second die, this
being overtopped by a statue eight feet four inches
high, representing a volunteer soldier, with musket at
parade rest. The base is of Rockport granite and the
rest of Italian marble, and the whole is enclosed by an
iron fence. Chiseled upon the tablets are the names of
those who fell in the conflict, accompanied by the following inscription: " In grateful tribute to the memory
of those who, on land and on the sea, died that the Re-
public might live, this monument is erected by the citizens of Haverhill, A. D. 1869."

Haverhill has had more than one opportunity to
prove itself superior to severe calamity in the shape of
fire. In 1775, just at the outbreak of hostilities between the Colonies and Great Britain, a fire occurred,
which, spoken of by them as the " late dreadful fire in
this town," was enough, with other causes, to detain at
home the Haverhill delegates to the Provincial Congress. It destroyed seventeen buildings, covering the
whole side of one of the main streets, and would doubtless rank, in point of proportionate importance, with
some of the later fires, such as, for example, one that
occurred in 1873, which " burned out " thirty-five business firms, caused the loss of two lives and destruction
of one hundred and fifty thousand dollars' worth of
property, and which was only extinguished by aid from

abroad. This was looked upon at the time as the worst fire in Haverhill's history, but it was dwarfed into insignificance by the "great fire" of the spring of 1882, and which is noteworthy, not alone or chiefly for the suddenness of the calamity or the magnitude of the loss or the completeness of the disaster, sudden and great and complete as these undoubtedly were, but rather for the abounding energy, determination, and speed with which the even then smoking ruins were removed, and replaced by structures far better than the original.

At twenty minutes before twelve o'clock on the night of Friday, Feb. 17, a fire was discovered in a wooden block among the shoe manufactories, which, it is agreed, a few pailfuls of water could at first have put out, but which spread with such amazing and, as it were, virulent rapidity, that the fire department, though promptly on the spot and working with the intensest energy, soon recognized its powerlessness to cope with the flames. Telegrams were sent to other cities, near and remote, for help, and very opportune and valuable aid was rendered by the departments of Newburyport and Lawrence. Had it not been for this, it is probable, that the fire, which, as it was, was confined chiefly to the shoe manufactories, would have spread to the retail stores and the dwelling-houses of the city, and, in fact, that its ultimate limit would have been a mere matter of chance. As it was, however, it was only with the utmost difficulty that help was obtained. The telegram sent to Boston was not delivered. The steamer was got from Lawrence only by the exertions

of the general ticket agent of the railway, who broke
open the railway telegraph office at Haverhill and
thence sent the necessary orders to the employes of
the railway at Lawrence. At Newburyport, the platform cars were frozen on the track, and it was with
great difficulty that the steamer was finally got under
way. It was only with the severest and most painful
efforts that the fire was at length controlled. It was
bitter winter weather, and there were those among the
most exposed of the firemen who lay in water several
inches deep, their clothes frozen so stiff that they were
unable to move except as rolled over by their companions, in order to direct a stream upon an important point.
It is worth while to say here, that it was this fire that
called attention to the need of an increased water supply in case of fire. Had the present abundant high-
pressure service then existed, it is safe to say that "the
Haverhill fire" would not have been.

The sun of Saturday morning shown upon the
ruins of two million dollars' worth of property, including one savings and two national banks. About three
hundred firms and individuals, engaged in various sorts
of business but chiefly shoe manufacturing and collat-
eral branches, were "burned out." One man was
killed during the fire, and another severely injured.
Live cinders were blown four miles off; the light of
the fire was seen in Boston, thirty miles distant; and
the sky all around was so brilliantly illumined by the
fire that a newspaper was read by its light at George-
town, six miles away. The fire not only destroyed
nearly every factory in the "shoe district" and thus

threatened to blot out the chief industry of the city, but
it burned as well the machinery, lasts, dies, patterns,
samples, and trimmings that were in readiness for the
large orders for which customers were already waiting.
In the face of the emergency, however, the chief losers
rose to the occasion, and, though great inducements
were held out to them by other towns and cities to
locate elsewhere, not one of them did so.   One or two
left the city, but only for a short time.

The first news the owner of the only building
spared by the flames (then absent in Washington) had
of the occurrence of the fire was contained in half a
dozen telegrams sent by men who wanted to rent his
unoccupied space and sent before their own walls had
fallen in.   At four o'clock on Saturday morning, while
the fire was still burning, the president of the First
National Bank called a meeting of the directors, which
was held at nine o'clock, when it was voted to rebuild
at once, a committee was appointed, and the plans were
well under way before night.   By the next Monday
nearly one half of the burned-out firms had secured
places and were employed in taking orders and preparing for the renewal of business, scattered in various
parts of the city, in attics, barns, sheds, dwelling-houses,
and abandoned buildings.   By the same Monday night
one prominent leather house had sold thirteen thousand
dollars' worth of leather for immediate use by manufacturers of the burned district.   The later region
presented a picturesque appearance, its ragged heaps of
bricks and stone dotted with signs announcing removals
to more convenient quarters.   In three days one half of

the firms had started their machinery. The workmen
had been already paid off; in a week the fire was a
thing of the past, and in a month everybody was settled
and looking forward only to the time when the work of
rebuilding should be finished. On the Tuesday after
the fire two cases of shoes were shipped by one of the
burned-out firms; and on Thursday, while the fire was
still smoking, the first brick was laid for a new building
in the burned district, where thirteen millions were to
be used before the mason laid aside his trowel. In eight
days a wooden building had been put up, and its upper
story got in readiness for the shoe-stitching firm that
had leased it.

The operatives lost, of course, all their tools; and
destitution and suffering would have been prevalent but
for the immediate formation of a relief committee,
which distributed the funds raised by the citizens and
the very handsome gifts received from abroad, — from
former residents of the city, including the poet Whittier, and from the large customers of the burned-out
firms. It should be stated that a large proportion of
the contributed funds found no use and was returned to
the donors. The fire was, in the nature of things, a
terrible shock to the community; and it was naturally
feared that it was a shock from which the city would
not recover and that it would cause a permanent
paralysis of the industry to which it owed its growth
and prosperity and in which all its hope for the future
rested. But the very greatness of the shock seemed to
produce an intense reaction, and the prevailing expression was one of hope and buoyancy. To quote a

recent writer. " Business soon became active again, and
the object of the sufferers was to resume operations in
the old localities as soon as possible. This was largely
accomplished before the first anniversary of the fire,
and in a most satisfactory manner. Beautiful and substantial buildings had been erected in place of those
destroyed, and the anniversary of the outbreak was
celebrated by a spirited banquet. Through the exhibition of pluck and energy made by the sufferers,
they won the sympathy of the entire business community of the country. The fire, distressing as it
seemed, is generally admitted to have been a blessing
in disguise."

There have been occasional fires since, some of
which threatened great destruction, and two of which
compelled aid to be sought from other cities. Not the
least serious was the one that destroyed the city hall a
little before noon on Tuesday, Nov. 6, 1888, causing a
loss of about forty thousand dollars. The fire caught in
the attic from an unknown cause, and burned with such
remarkable intensity and rapidity that no efforts of the
fire department availed to check it, and it continued until the roof had fallen in, with the clock-tower and bell,
the whole interior of the building destroyed, and only
the blackened walls left standing. During the fire a
number of sparks were carried, by the strong southwest
wind that was blowing, upon the roofs of buildings on
the eastern side of Main Street, some of which suffered
damage. The Center church sustained the severest
loss. A disastrous conflagration was at one time
threatened but was averted. The city hall was

CITY HALL, BURNED NOV. 6, 1888.

erected in 1861 on the site of the old Town hall which
it replaced.   It was a massive three-story structure of
brick ornamented with freestone, one hundred and fifteen feet long, sixty-seven and a half feet wide, with a
clock-tower on the front eighteen feet square.   The
work of restoration was not long delayed, and from the
ruins has already arisen a new structure, with a better
tower, a larger and finer bell, and an illuminated clock,
and which bids fair to excel the one destroyed.

WITHIN AND WITHOUT.

———————

Haverhill's situation and natural advantages have been remarked upon from the earliest times, and have amply certified to the acumen of Ward and his associates of 1640 when they chose this spot for their plantation. In the first place, the river that edges it is one of the most noteworthy of ancient or modern passageways to the sea. It turns more spindles than any other river, being the most noted water-power stream in the world, seventy-eight thousand six hundred horse powers being utilized in 1880 on the Merrimack and its tributaries, probably a greater extent of occupied water-power privilege than on any other drainage basin of the same size in America. The total fall of the river is not great, but it is concentrated at a few places, thus occasioning its wonderful adaptedness to be utilized as motive power. Having its source up in the impenetrable fastnesses of the White Mountain wilderness, fed by the inexhaustible outpour of the beautiful Lake Winnipesaukee, it sweeps by the mills of Manchester,

Nashua, Lowell, and Lawrence, until at Haverhill, sixteen miles from its mouth, it begins to smack of the sea, since here is the head of navigation and here the tide rises and falls.

Haverhill lies on the northern edge of Essex County (itself the northeastern corner of Massachusetts), on the northern bank of the Merrimack River, and is one of the chief stations on the Boston and Maine railway. It is thirty miles from Boston on the highway and thirty-three by rail, while it is eighty-three miles from Portland, Me., the eastern terminus of the main line of the railway, and ten miles less as one drives over the road. It is nine miles distant from Lawrence, fourteen from Newburyport, eighteen from Lowell, twenty-two from Salem, and thirty from Portsmouth, N. H. It is not only one of the most important places on the main line of the Boston and Maine system, but, by a branch running through central Essex, it has free communication with the inland county towns, with Newburyport, and with the whole eastern division of the Boston and Maine. Three highway bridges span the river at Haverhill and connect with it Bradford, Groveland, and West Newbury. The river plays no unimportant part in its affairs, since it affords the opportunity for delightful recreation in the season, the means of cheap freightage for bulky articles, and a continual means of escape for the city's sewage. It is not so much a channel of commerce as it was in the elder days, before the railway had been heard of and when the shipwright's hammer and the calker's tool still rang frequent in the Haverhill yards. The first steamboat, in fact, that ever floated on

BOSTON AND MAINE R. R. BRIDGE.

the Merrimack was built in Haverhill in 1828. The chief obstacles in the way of river commerce above Haverhill are the shoals and rapids that intervene between it and Lawrence. Attempts have been made by the national government to deepen and widen the channel, and some coal lighters have been towed to Lawrence and small steam vessels of light draught have even ascended the river to that point since the dam was built at Lawrence, before which time steamers plied between Lowell and the ocean; but the work has been given over, at least for the present. At Haverhill, however, the river has a width of six hundred feet and a channel depth of eight feet at high water, and vessels of two hundred tons come up from the mouth of the river to lie at the Haverhill wharves, laden with lumber, stone, and coal. In the summer time, pleasure steamboats ply up and down the stream and convey thousands of passengers by a delightful voyage to the beaches at the mouth of the river.

Not far below Haverhill Bridge is a long but rather narrow island, opposite the establishment of Col. Harry H. Hale on the Bradford side of the river, of which estate it forms a part and to which it has given the name of "Island Stock Farm." It is utilized for pasturage, and a half-mile track has been made there in which to exercise Col. Hale's colts.

The city is nine and a half miles long, with an average width of three miles, extending over twenty-four square miles. There are one hundred miles of streets, twenty-seven miles of sidewalks, fifteen miles of sewers. The disproportion between the highways and

HAVERHILL BRIDGE.

such adjuncts as sidewalks arises from the extensive territory outside of the city proper, which is highly
productive and for the most part highly cultivated.
The city is traversed by three small streams, tributaries
of the Merrimack, two of which have been utilized for
grist-mills and saw-mills, while the remaining one
affords enough water-power to turn the wheels of a
large flannel mill.   Haverhill is noteworthy for the fact
that there are four ponds within its limits, and three of
them within a mile of the city hall and within a half
mile of each other.   All four of them are now used to
supply the city with water for drinking purposes.   They
are valuable, however, not alone for the abundance with
which they administer to the thirst of the city, or for
the ice which makes more endurable the summer heat
or affords a smooth surface to the swift foot of the
wintry skater, but also as adding a variety, a pictur-
esqueness, and a charm to the landscape such as few
cities can boast.   The smallest of the four covers about
thirty-eight acres and was the first used for aqueduct
purposes because it appears to be fed entirely by
springs.   The next in size, covering but two or three
acres more, supplied the head for the first mill-powers
utilized in the town.   The other two are much larger,
one of them, its waters remarkably clear and trans-
parent, occupying an area of one hundred and seventy-
five acres, while the largest of all, Lake Kenoza (lake
of the pickerel), includes two hundred and forty acres.
It is fifty feet in depth in some places, and, though but
a mile from the city hall, is picturesquely surrounded.
It once abounded in pickerel, and through its outlet

ISLAND STOCK FARM.

alewives and salmon used to crowd in spawning time.
The woods on its edge were long the haunt of several
species of game and were therefore very attractive to
the sportsman. It still affords to the residents of the
city, as it long has afforded, a pleasant resort, within a
a convenient distance, for parties of pleasure, who
doubtless often find expressive of their own feelings the
words that the poet Whittier, himself from boyhood
familiar with its shores, sent to its christening, —

" Kenoza! o'er no sweeter lake
 Shall morning break or noon-cloud sail, —
No fairer face than thine shall take
 The sunset's golden veil.

" Long be it ere the tide of trade
 Shall break with harsh-resounding din
The quiet of thy banks of shade
 And hills that fold thee in.

" Still let thy woodlands hide the hare,
 The shy loon sound his trumpet note,
Wing-weary from his fields of air,
 The wild goose on thee float.

" Thy peace rebuke our feverish stir,
 Thy beauty our deforming strife;
Thy woods and waters minister
 The healing of their life."

The older and more compact part of the city lies
along a southward-looking slope that rises sharply

from the river, and its houses, at first closely clustered
for neighborhood defence in Indian times, now stretch
for miles up and down the stream. It is not unlikely
that the natural beauty of their clearing soon caught the
eye of the early settlers, and that they set their houses
away up on the bank, the road running in front of them
and thus separating them from the river, with the
intent to allow no buildings on the opposite side and
thus insure to them on their high land an unobstructed
view of the stream. It was almost inevitable, however,
that the growing value of the riparian land should
compel its utilization; and the river road of the settlers
has become the main business street of the city, closely
built on each side with shops and stores in the region
of trade, wharfage occupying the rear of the riverward
side.

The general surface of the city is undulating,
though some of the ascents and descents to and from
the river are quite sharp. There is little or nothing,
even in the outlying districts, of the precipitous sides
and jagged tops that are not uncommon features of our
New England river towns, but the eminences are in
general not very difficult of ascent, rounded, and often
cultivated to the top. They are noteworthy, too, for
being detached summits, instead of being continuous
upland or chains of hills, thus affording a greater
variety to the landscape, and suggesting, as the city
grew, fit spots for the erection of more pretentious and
more costly residences, in keeping with the increasing
wealth and enterprise of the city. Many of the hills
have already been utilized for this purpose, some of

the nearer slopes being more or less closely occupied
by types of the modern handsome house, and many
acres of land have thus been brought to a present or
prospective market. Whether one prefer the outlook
on river, lake, or meadow, there is no lack of eligible
building sites, not far removed from the more compact
city. Close to the river, even, rise several eminences,
one to the east and one to the west of the city proper,
each of which affords from its summit a beautiful view
of the Merrimack flowing at its feet and of the
towns beyond. They bear the somewhat curiously
antithetic names of Golden and Silver, named, however, not from any metallic properties, actual or
metaphorical, but from some early and long forgotten
owners. Washington, in his tour of New England in
1789, passed through Haverhill, and his admiration of
the beauty of its situation has been sedulously preserved in tradition and has been set to verse by
Whittier, himself an ardent lover of thecharms of his
native town.

> "Midway, where the plane-tree's shadow
>   Deepest fell, his rein he drew:
>   On his stately head, uncovered,
>   Cool and soft the west wind blew.
>
> "And he stood up in his stirrups,
>   Looking up and looking down
>   On the hills of Gold and Silver
>   Rimming round the little town, —

" On the river, full of sunshine,
     To the lap of greenest vales
Winding down from wooded headlands,
     Willow-skirted, white with sails.

" And he said, the landscape sweeping
     Slowly with his ungloved hand,
' I have seen no prospect fairer
     In this goodly eastern land.' "

About a mile from Kenoza Lake rises an eminence
known by the name of Great Hill and which is the
highest land in the town. It is three hundred and
thirty-nine feet above the ocean and is the second
highest elevation in Essex County. "The view from
the summit of this hill," writes a local historian, "is
the most extensive and interesting of the many similar
views to be obtained in the town. Portions of more
than twenty towns in Massachusetts, and nearly or
quite as many in New Hampshire, are easily distinguished by the naked eye. To the east stretches the
broad Atlantic, whose deep blue waters, dotted with
the white wings of commerce, are plainly seen, from
the Great Boar's Head to Cape Ann. Near its edge,
and partially hidden from our sight by Pipestave Hill
in Newbury, are seen the spires and many of the
houses of the city of Newburyport. To the right, the
eye can distinctly trace the outline of Cape Ann from
Castle Neck to Halibut Point. With the aid of a glass
several villages upon the Cape are made visible. As
we sweep around from east to south, nearly all the

most prominent hills in northern Essex can be distinctly seen and easily identified. To the south and southwest, portions of the villages of Groveland, Bradford, Haverhill, North Andover, Andover, and Methuen, and the city of Lawrence, can be seen, peeping above the intervening hills. To the southwest, the Wachusett; to the west, the Monadnock; and to the north, the Deerfield mountains are easily distinguished. To the northwest, the village of Atkinson, with its celebrated academy, is spread out in bold relief. To the northeast is seen the top of Powow Hill, in Salisbury, so named from its having been the place selected by the Indians for their great "pow-wows," long before a white man gazed upon the waters of the Merrimack from its summit. Turning again to the south, we notice, almost at our feet, the beautiful Lake Kenoza, glistening in the sun like a diamond encompassed by emeralds. Once viewed, the memory of this lovely landscape scene will never be effaced, —

————

'the faithful sight
Engraves the image with a beam of light.'"

————

In fact, in nearly every part of the city are hills of more or less prominence, some of the remoter ones still affording pasturage for cattle, while on the southward-looking slopes of others the grape mellows in the

autumn sun. On a great rock at the summit of one of
them, bearing the unique and perhaps inexplicable
name of Brandy Brow, four towns meet cornerwise, —
two, Plaistow and Newton, in New Hampshire, and
two, Haverhill and Amesbury, in Massachusetts. Another overlooks the humble birthplace of the poet
Whittier, the Mecca of so many travelers' feet, while
from other hills in the eastern parish may be had a
fine view of the Merrimack and of the wide-stretching
East Meadows, by which the early townsmen set so
much store. Everywhere broken, offering glimpses
now of pond and now of river, affording a wider outlook upon more distant scenes at every turn, nothing
" can stale the infinite variety " of the landscape.

Schools.

The riches of the Commonwealth
Are free, strong minds, and hearts of health;
And more to her than gold or grain
The cunning hand and cultured brain.

I.

There rises before one at the moment of beginning this sketch of the schools of Haverhill two pictures, — the one dim, imperfect, its features almost obliterated by the passing years, a canvas where a few, silent, enshadowed figures are faintly seen; the other bright with strong, fresh colors, sparkling with life, thronged with faces as the paintings of Raphael are with angel heads; the one, that first school of Haverhill taught by Thomas Wasse for ten pounds a year, its place of meeting some private house, whither by rude cart-paths or footways, through woods where beasts or savage

Indians lurked. the few children of the rude settlement
of two hundred years ago went to be taught to read
and write and cast accounts; the contrasting picture,
the attractive spacious school-rooms, fitted with all
that ingenuity can suggest for comfort or for teaching,
wherein the present generation of children gathers to
be taught. in ways and with helps of which the rude
forefathers never dreamed. the knowledge and wisdom
of to-day.   There lies before the writer a volume containing the Haverhill school reports of many years.
and. as one reads backwards through these, and.
beyond them. through the fragmentary and far separa-
ted sketches of the schools of ancient days. one cannot
but recognize with what faith and deeds the valiant-
souled and earnest-hearted fathers of the town sowed
the seed which has grown into the magnificent school
system of which we are justly proud.

It should not be forgotten. that those noble men
who came to New England in 1630 and the years
following. men "who." Macaulay says. "forever illus-
trious in history. were the founders of the Commonwealth of Massachusetts." were neither adventurers
nor untaught dissenters. They were many of them
university men. They brought with them their well
selected libraries. They brought. also. the belief that
the education of the people ought to be the first concern of the state. Their judgment of what that education should be was no narrow and merely utilitarian
one. They took as the guiding purpose of their action
the same broad idea that formed but lately the key-
note of the address of the orator at the dedication of

the Haverhill High School building: "In the matter of education the natural flow is from the heights to the plain. * * * * There must be elevated fountains of knowledge in order that these blessings may be generally distributed among the common people." "Probably," says the historian of American literature, "no other community of pioneers ever so honored study, so reverenced the symbols of learning; theirs was a social structure with its corner-stone resting on a book."

The first public school established was the Boston Latin School. This school, founded so much earlier than Harvard College that it is said to have "dandled Harvard College on its knees," owed its existence largely to two men, the far-seeing governor, Winthrop, who knew that ignorance was the "darkest lair of Satan," and the Reverend John Cotton, "to whom," Dr. Increase Mather says, "New England oweth its name and being more than to any other person in the world." Cotton was a graduate of Trinity College, a fellow of Emmanuel College, a man recognized in England as of great ability and learning, and in New England the acknowledged center of vast influence in church and civil affairs. All that was precious to him in his memories of England he transplanted to America. "When he saw the children growing up he thought of the school, the *free* school, to which all could go; and with his own love for classical literature, and his partiality for the privileges of a collegiate education, the memory of a free grammar school where Greek and Latin were taught may have risen to his mind, and he may have said, "Here, also, where the trees of the forest are not yet felled and the wild Indian is at our doors, here let such a school be established, free for all. And let this one be the forerunner of a thousand more that shall follow."

By the influence of such men in 1647 the General Court passed the following law, "in order that learning may not be buried in the graves of our fathers:" "It is, therefore, ordered that every township in the district, after the Lord hath increased them to the number of fifty householders, shall then forthwith appoint one within their town to teach all such children as shall resort to him, to write and read * * * * * and it is further ordered, that, when any town shall increase to the number of one hundred householders, they shall set up a grammar school, the master thereof being able to instruct youth so far as they may be fitted for the university * * *."

From this influence and this order came the public schools of New England.

"Yet with our fathers we are one
    At heart, whatever change betide;
Still shines for us their tireless sun;
    Their truth still waits us for our guide."

II.

The larger settlements, like Boston and Salem, did not, however, contain all the men of education and high purpose. In the little frontier town of Pentucket, afterwards Haverhill, the minister, John Ward, was a man "learned, ingenious, and religious, — an exact grammarian, and an expert physician," — a Master of Arts of the University of Cambridge, England. The few men associated with him in founding this settlement, and who lovingly and reverently called him *Teacher*, though not as well educated as himself, were by no means illiterate. There was no schoolmaster chosen for fourteen years after the order of the General Court, but the colony did not until that time reach the required number of householders. Moreover, by reason of its being a frontier town, it had more difficulties with which to contend than the other settlements. The Ipswich father of that day had to accompany his children to the school to guard them from the wolves. The Haverhill father must fear the wily Indian as well as the forest beasts. Though there be no historical record to confirm it, one must believe that the children of the colony were taught at home until the first master was chosen; that, amid the labors and watches of the day or by the glowing pine knot at night, the father gave to his sons what knowledge he himself held. The town records of the earlier years make frequent mention of schools, now the authorizing of the hiring of Thomas Wasse as schoolmaster at ten pounds a year, later the raising of thirty pounds for school purposes, again the engagement in 1702 of a Mr. Tufts for a salary of thirty-four pounds, but in 1703 the town voted "that, on consideration of their troubles with the Indians, nothing should be done about getting a schoolmaster," and in 1705 the General Court, because of their impoverishment by the Indian war, excused all towns of less than two hundred families from observance of the school law for three years.

It may seem unbefitting a volume of this kind to make the sketch of the schools at all historical, but a view of the education of the past is useful not only as a contrast with that of the present in the material equipment, but as showing that the love of learning and the high aims of our schools are deeply rooted in the past. We have no more solicitude for learning than they had in those early days, when the New England matron said to her son, "Child, if God make thee a good Christian and a *good scholar*, thou hast all that thy mother ever asked for thee."

It would be of little value here to note the varying fortunes of the schools in the past century, but it is interesting to note that a hundred years ago, in 1789, the first school regulations were adopted by the school committee of Haverhill. Although new methods of teaching have replaced the old, we must recognize, as we read some of these century-old rules, that the purposes of knowledge remain unchanged. Indeed, with scarcely the modification of a sentence, we might place in our regulations these framed a hundred years ago: That " the master consider himself as in the place of a parent to the children under his care, and en-

deavor to convince them by mild treatment that he
feels a parental affection for them; that he be sparing
as to threatenings or promises, but punctual in the execution of the one and the performance of the other;
that he never make dismission from school at an earlier
hour than usual a reward for attention or diligence, but
endeavor to lead them to consider being at school a
privilege, and dismission from it a punishment; that
when circumstances admit he suspend inflicting punishment until some time after the offence is committed;
that he impress upon their minds their duty to their
parents and masters; the beauty and excellence of
truth, justice, and mutual love; tenderness to brute
creatures, and the sinfulness of tormenting them and
wantonly destroying their lives; the duty which they
owe to their country and the necessity of a strict obedience to its laws; and that he caution them against the
prevailing vices, such as Sabbath-breaking, profane
cursing and swearing, gaming, idleness, etc."

Books have changed and will change, and sciences
and studies and methods of interpretation, but the pupils of a hundred years ago were taught as the pupils
of to-day are taught, and the pupils of a hundred years
hence shall be taught and trained, " in the purposes of
knowledge, in the love of justice and generosity and
patriotism, in respect for themselves, and in obedience
to authority, and honor for man and reverence for
God."

Though we live when liberty is larger and civilization richer and humanity more tender, we cannot afford to despise or overlook the foundations that were

so deeply and strongly laid in the past that we can
safely rear thereon broadly and high, to-day, our institutions.   In education the objects to be achieved alone
are stable; the methods must vary with the varying intellectual surroundings and demands of the age and the
generation.   What Emerson calls the "work of
divine men," "to help the young souls, add energy,
inspire hope, and blow the coals into a useful flame,"
is shown to have been the guiding moral purpose of
the first regulations of the Haverhill schools of one
hundred years ago, and is to-day the one purpose of our
more ambitious system of education.   Side by side
with the training that shall cultivate the power of
thinking, give knowledge, promote loyalty, and industry, and high ambition, we seek to place the inspiration to truthfulness, purity, and courtesy.

III.

The schools of Haverhill to-day stand abreast with
the best in the country.   Sufficiently progressive to
adopt whatever is an improvement upon previous
methods, sufficiently conservative not to be swept along
by every new fashion in education, making a specialty
of no one branch of the school curriculum, the schools
furnish, from the lowest primary grade to the highest
high school grade, a course of study that seeks the
symmetrical and progressive development of the child.
    The school board, of which the mayor is *ex-
officio* chairman, consists of eighteen members, one

being chosen each year from each ward and the term
of office being three years.

Beside the various sub-committees on the several
schools, there are standing committees on school-
houses, salaries, truancy, music, private schools, text-
books, and examination of teachers, and a prudential
committee for the examination of all bills against the
school department, their approval being necessary before the bill can be paid. The general board meets on
the third Wednesday of every month for the consideration of the school interests, and the prudential committee on the Monday preceding the meeting of the
board.

Happily the election of school committee has been
determined by fitness instead of political questions,
and the board, while differing occasionally, as honest
men may, about methods, has been unanimous in seek-
ing to obtain and maintain the best schools possible.
While keeping a strict watch to check any extrava-
gance or needless expenditure of money, it believes
that the first element of economy is efficiency. The
teachers are elected annually in June, at which time
such changes or dismissals are made as seem necessary. In the selection of teachers favoritism and personal desires are not factors, the qualifications of the
applicant in respect to character, education, and the
power to teach being alone considered.

The superintendent of schools is the secretary of
the board. He keeps the records, buys all school supplies and distributes them to the schools, makes out the
weekly pay-rolls, and arranges and presents all bills to

the prudential committee. By a system of monthly reports from each school, he is able to present each month to the school board the exact condition of the schools, and to show wherein there is improvement or need of improvement. As superintendent of schools he conducts examinations, has charge of promotions, visits each school, and advises with the several sub-committees upon questions of changes in course of study, text-books, discipline, etc. He keeps watch to know what progress or changes other places are making in methods of education, and is in all matters the executive agent of the board. Every month the teachers of each grade meet with him for comparison, discussion, and suggestion, and thereby an *esprit de corps* of great value is maintained. In the grammar schools the principal, under the direction of the super-intendent, supervises carefully the work of each grade in his own building. The principals of all the schools meet at intervals with the superintendent to discuss school interests and obtain uniformity of methods. The object of this arrangement of school supervision is to obtain in each school the best results, but, while the system is made as complete as possible, there is sufficient elasticity to allow of individual work by the teachers and individual training of the scholars.

The course of study is so arranged that each branch shall receive its own proportional amount of time and attention. In reading, ease, fluency, and expression are sought; and each lesson is preceded by a vocal drill to obtain clear enunciation and variety in expression. In writing, a regular drill is given, to ob-

tain an easy control of the muscles of the arm and the
fingers. In geography and history, the scholars are led
to read widely, to compare authors, and to study by
topics the countries or the epochs. The study of language begins with the child's entrance to school and
continues through the full course. The course in
drawing has just been re-arranged in order to make it
a progressive study of form and objects through all the
years. The music is under the direction of a special
teacher.

Promotions of classes are made yearly, and are so
arranged as to prevent as far as possible any nervous
and unnatural strain upon the child, the estimate of the
teacher under whom the pupil has been during the
year being the especial basis of promotion. Written
tests and exercises are given frequently to cultivate ex-
actness and power of expression, and to show what
subjects need reviewing. In all promotions the individual child is considered, and the question asked,
" Is it best for him to go on or to review the work?"
The school session is freed from all tediousness by
numerous changes, and by the introduction of suitable
gymnastic exercises. For some years no out-door recess has been given. This no-recess plan has been a
feature of the school system long enough for an un-
prejudiced judgment to be formed of its results. It is
found that it is much easier to maintain school discipline, and that there is much less opportunity for the
forming of evil habits or associations under this than
under the old system, while the shorter school session,
the short in-door recess, and the ready permission to

leave the room when necessary prevent any extra fatigue and any injury to the health.

Entering the lowest primary grade, the child comes immediately under the care of teachers chosen because of their especial fitness for primary work. From his very entrance into school, he is trained to read, to write, to measure, to observe; he is taught the correct use of language, and is led to express his thoughts in complete sentences; cleanliness, order, and courtesy become as habits to him, while, so far as the influence of the school-room extends, he is restrained from cruelty and coarseness and the more flagrant vices.

The evening schools are open for twenty weeks, three evenings a week, and in them the division is into small classes, each having a separate teacher, in order that much individual work may be done. There is an evening school of mechanical drawing, and one of freehand drawing, and a school for instruction in bookkeeping, in addition to separate schools for the instruction of males and females in the ordinary grammar-school branches. In these schools the city gives most willingly not only what the state demands, but what contributes to the advancement of those who, debarred by the necessity of labor from the day schools, desire to obtain an education.

There are eighty public schools in the city, occupying twenty-three buildings, and taught by ninety-three teachers. The number of pupils in the public schools is about 3,000; in parochial schools 1,000; in other private schools 50. The city spends annually for the support of its schools about $65,000. In 1886 it ex-

HIGH SCHOOL.

pended more money per pupil than any other city in Essex County, and was outranked in the state only by Boston, Newton, New Bedford, Somerville, and Cambridge, none of which are purely manufacturing cities. In proportion to its valuation it expends more than any city in Massachusetts save Gloucester. In this comparison towns are not included.

The city furnishes free to all pupils all books, slates, stationery, etc., used in the schools, and offers to the children of rich and poor alike the best teaching that it can obtain, the best courses of study that it can devise, the best text-books and the most complete aids for study, during a school course of thirteen years, carrying the student to the very doors of the scientific or academic university, and all without the expenditure of a dollar.

The High School is beautifully situated on a commanding site on Crescent Place fronting a small park, and occupies the place where Harriet Newell, one of the first missionaries of the American Board, was born, as well as the place where stood the Center school, the first and for many years the only grammar school of the town. The architecture is Roman and Grecian combined, and freed from all the trickeries of form and ornament, with its simple lines and true proportions, is of great dignity and beauty. The building is three stories high above a granite basement and is handsomely built of brick with sandstone trimmings. The basement contains, in addition to the most excellent sanitary arrangements and the boilers for the steam-heating aparatus, a chemical laboratory fitted with

desks and furnished completely for experimental
study, and a philosophical lecture-room, both large
and well ventilated. Above, on the first floor are the
spacious school and recitation rooms, the rooms of the
school commitee and the office of the superintendent
of schools. The second floor contains, in addition to
the school and recitation rooms, the school library and
the office of the principal. The third floor contains
the large school hall where the school assembles for
devotional exercises, for music, and for public declam-
atory exercises. It contains also two rooms fitted for
the teaching of instrumental and free-hand drawing,
and containing a large number of casts and studies.
An arrangement of gaslights and screens gives facilities for the study of light and shade effects. The
corridors are high and wide, the staircases of easy
ascent, the cloak-rooms and teachers' apartments light
and ample. Electric bells and speaking-tubes com-
municate with the principal's room from all parts of
the building, and the edifice, first occupied in 1874,
and costing with the lot about $110,000, is a model of
comfort and convenience. From its upper windows a
large portion of the city may be seen, and the windings
of the beautiful Merrimack traced for a long distance.
The halls and school-rooms are adorned with pictures
and busts, gifts from the Alumni Association and the
graduates and friends of the school. The Alumni
Association is one of the oldest of such institutions,
and perhaps the most prosperous. It gives two recep-
tions during the year, invitations to which are eagerly
sought, and it has a quite large fund safely invested,

the income of which has been devoted for some time to
the purchase of pictures for the beautifying of the
school-room walls.

The school has about 200 pupils. Its corps of
teachers is a master, two submasters, four female
assistants, and the instructor in music. The most of
these teachers have been long connected with the
school, and all have especial fitness for the departments
of instruction under their charge. The Master is a
graduate of Harvard College, the first submaster of
Dartmouth, the second submaster of Brown. There
are three full courses of study, each of four years, the
Classical, the English and Classical, and the English.
The traditions of the school are of high scholarship,
and it is the constant aim of the officials to use the best
methods and secure the best results. It has been the
pride of the school to enter its sons at Harvard or
Dartmouth or Williams or Amherst as well trained as
the boys from Exeter or Andover, to place those who
choose a scientific course in the Institute of Technol-
ogy unconditioned, and to present its daughters fully
prepared for the examinations at Wellesley, Smith, or
the Harvard Annex. How intimately it is connected
with the civic and social life of the place may be seen
in the fact that among its former pupils are the mayor
of the city, its civil engineer, many of its bank cashiers
and tellers, several of the trustees of the Public Library,
the superintendent of schools, the master of the High
School, and the majority of the public teachers, members of the school board, and very many of those who,
in the various literary clubs of Haverhill, promote the

social and literary interests of the city. Among those who have gone forth from this to other fields of labor and usefulness, and whom the High School has trained and prepared, are those who fill all grades of honor and of trust,—the president of the national senate, lawyers and preachers, scientists and business men.

But be the power and success of the school shown in the lives of those who serve in more important or more humble offices, the school seeks always to leave those who go forth from it more mighty in mind, more mighty in heart, richer in the power of usefulness, to place them more surely under the guardianship of " the three great angels of Conduct, of Toil, and of Thought."

The list of the present corps of teachers, and the course of study are appended:

Clarence E. Kelley, A. M., Harvard '73, Master.

James D. Horne, A. B., Dartmouth '84, Submaster.

Walter O. Cartwright, A. B., Brown, '81, Submaster.

Harriet O. Nelson, English Literature and Latin.

Mary S. Bartlett, Latin and Physiology.

Nellie M. Moore, French and History.

Mira W. Bartlett, Geometry, Drawing, and Botany.

W. W. Keays, Instructor in Music.

HAVERHILL HIGH SCHOOL.

COURSE OF STUDY.

ENGLISH COURSE.

FIRST YEAR.

First Term.	Second Term.
Algebra.	Algebra.
English History.	French History.
Book-keeping.	Book-keeping.
Civil Government.	English.
Drawing.	Drawing.

SECOND YEAR.

First Term.	Second Term.
Geometry.	Geometry.
Physiology.	Botany.
Arithmetic.	Arithmetic.
English.	English.
Drawing.	Drawing.

THIRD YEAR.

First Term.	Second Term.
French.	French.
Rhetoric.	Roman History.
Physics.	Rhetoric.
Greek History.	Physics.

FOURTH YEAR.

First Term.	Second Term.
French.	French.
English Literature.	English Literature.
Chemistry.	Astronomy.

ENGLISH AND CLASSICAL COURSE.

FIRST YEAR.

First Term.	Second Term.
Algebra.	Algebra.
English History.	French History.
Latin Grammar and Reader, Latin Composition.	Latin Grammar and Reader, Latin Composition.
Civil Government.	English.

SECOND YEAR.

First Term.	Second Term.
Geometry.	Geometry.
Physiology.	Botany.
Cæsar, Latin Composition.	Cicero's Orations.
English.	English.

THIRD YEAR.

First Term.	Second Term.
French.	French.
Greek History.	Roman History.
Virgil.	Virgil.
Physics.	Physics.

FOURTH YEAR.

First Term.	Second Term.
French.	French.
English Literature.	English Literature.
Chemistry.	Astronomy.

CLASSICAL COURSE.

FIRST YEAR.

First Term.	Second Term.
Algebra.	Algebra.
English History.	French History.
Latin Grammar and Reader, Latin Composition.	Latin Grammar and Reader, Latin Composition.
Civil Government.	English.

SECOND YEAR.

First Term.	Second Term.
Geometry.	Geometry.
Greek Grammar and Reader.	Greek Grammar and Reader.
Cæsar, and Latin Composition, Sight Latin.	Cicero's Orations.
English.	English.

THIRD YEAR.

First Term.	Second Term.
Algebra.	Algebra.
Xenophon, Greek Composition.	Xenophon, Greek Composition.
Greek History.	Roman History.
Virgil, Sight Latin, Latin Composition.	Virgil, Sight Latin, Latin Composition.
Physics.	Physics and Astronomy.

FOURTH YEAR.

First Term.	Second Term.
French, English Literature, Geometry.	French, English Literature, Geometry.
Greek.	Greek.
Latin.	Latin.

GENERAL EXERCISES.

Compositions by all pupils. Vocal Music each week.
Declamations by boys.

A few rods west of the new High School building
stands the "outgrown shell,"—the old dwelling of
the school,—now occupied by the Whittier grammar
and primary schools. The seven or eight elms on the
beautiful lawn in front of the building may give the
pupils in the hot summer days a grateful idea of
academic shades and possibly the inscription High
School, still allowed to remain on the facade of the
building, may remind the pupils of what yet lies above
them. There is no building in the city around which
throng so many reminiscences. The land on which it
stands was given in 1826 as a site for an academy, and
the building was dedicated in 1827. The orator was
the Hon. Leverett Saltonstall of Salem, and the poet,
"a tall, slight, distinguished-looking but bashful youth
of nineteen, with strikingly beautiful eyes," was John

G. Whittier. who had just entered the school. Whittier was introduced to Miss Arethusa Hall, the preceptress. by the Hon. James H. Duncan as "a young man who at the shoemaker's bench often hammered out fine verses."

Fifty-seven years later a number of the surviving alumni of that old academy held a reunion in honor of Whittier, at which the beloved singer was present and for which he wrote a touching poem, —

1827 – 1885.

"The gulf of seven and fifty years
We stretch our welcoming hands across:
The distance but a pebble's toss
    Between us and our youth appears.

" For in life's school we linger on.
The remnant of a once full list.
Conning our lesson, undismissed.
    With faces to the setting sun.

*　　*　　*　　*

"The eyes grown dim to pleasant things
Have keener sight for bygone years.
And sweet and clear, in deafening ears,
    The bird that sang at morning sings."

The upper room, Academy Hall, was a place for lectures and balls and religious meetings, where "grave and gay alternate chased." The room below, the old school-room, has faintly echoed to the maiden "speaking" of some who afterwards won the applause of listening senates, and many who later spoke in the pulpit, on the platform, or at the bar. In 1841 the Academy became a High School. The building has been remodeled once or twice to suit the growing needs, and in 1869, at an expense of about $12,000, was extensively changed, while still keeping in its general external appearance the features of the old academy. The school has four teachers. The principal, Miss Sarah S. Noyes, though still on the sunny side of life, has taught in Haverhill schools for thirty years, and had a share in the training of many of the successive city governments, the school committee and the teachers. A short distance farther up Winter Street stands the Winter Street School building containing about 500 scholars, under the charge of twelve teachers, the principal being Mr. Charles W. Haley. This school is of high grade, and sends annually about forty pupils to the High School. The present building was built in 1856 and was dedicated with an address by the Hon. George S. Boutwell. It has since then undergone various changes to accommodate the growing school population.

The School Street Grammar School, under the charge of Mr. Fred Gowing, has about 300 pupils, with eight teachers. This school has been established for fifty years, although, like the other grammar schools, it has outgrown one dwelling after another during that time, and sent its overflow to other and newly created schools.

CURRIER SCHOOL.

There had long been a grammar school on Washington Street, a most delightful place when Haverhill was a village. In the stirring days of the rebellion its boys saw the sons of the village march past its gates on their way to the war; they saw the gallant hosts of Maine go by on the railroad just west; they saw also the home-coming of those who went forth, some with the cherished flag wrapped round their coffined forms, some marching beneath its stained and torn but yet victorious folds; and, through all the days of excitement, of grief, of waiting, of hoping, of victory, the nation's flag, made by the daughters of the school, floated from its upper window. The school — and the other grammar schools also — has its roll of honor, the list of its scholars who poured out their life-blood for the nation's defence. The Hon. George H. Carleton, the late mayor of the city, was its master in those days of action when its sons learned a practical lesson in patriotism. Later, trade invaded the quiet street, and tall brick buildings, bustling hives of industry, crowded back the quiet cottages, and made the removal of the school necessary. Following the "course of empire," Horace Greeley's advice, and the growth of the city, it went west, and on the fifth of June, 1873, occupied a new home on Silver Hill. The building was so superior to any other in the town that the school report says of it that its "prominence and superiority over all the surrounding structures is a correct indicator of the relative position which our educational system holds among the agencies of society as now constituted in our country."

BRADFORD ACADEMY.

As illustrating the growth of the city westward in the last fifteen years, it is interesting to note, that, when the Currier School was opened, it was current opinion that so large a building never could be used. It contained eight large school-rooms and a school hall. Three of these rooms were opened with an attendance of 198 scholars. Four rooms have since been added, and to-day twelve rooms are occupied, with an attendance of about 500, while two large brick primary schools of six and eight rooms respectively have been built in addition to accommodate that district.

The principal of this school is a woman, Miss Mary A. Tappan, who has been at the head of the school since the building was erected. It is possible that Haverhill recognizes the equality of the sexes more than any other city, for it pays the principal of this school the same salary that the male principals of the other grammar schools receive.

For some years now a training school for teachers has been in operation, and many of the most successful primary teachers are graduates of it. It is under the charge of a principal and an assistant principal. The number of pupil teachers is limited to sixteen. These must be graduates of the High School, or must successfully pass an examination upon prescribed subjects. The course of training is a year and a half, and the work is that of the four lower grades of the school course. The school has 200 scholars, and the pupil teachers, in additional to the theory of teaching and the normal work, are trained and tested by the care, the discipline, and the teaching of the four schools in the

CARLETON SCHOOL, BRADFORD.

building. The rank of this school is high, and applications to enter it come not only from the young ladies of the city, but from other cities and towns. The pupil teachers are subjected to constant examination, and to careful and kindly criticism, and receive certificates which state for what grade of teaching they are best adapted. Those who fail in the essential requisites of a teacher are, after careful trial, advised of their failure and quietly withdrawn from the school. The existence of this school shows the desire and care of Haverhill to obtain well trained and tested teachers for the youngest pupils.

Of the other grammar and primary schools it is needless to speak in detail. The same care, the same course of study, the same desire to do the best possible work is in them all. Sufficiently abundant in number and convenient in position to avoid large numbers or long distances, they leave no reasonable excuse for any child's not enjoying their privileges.

For those to whom private schools seem a necessity, Haverhill is most delightfully situated. There are private kindergarten schools for the youngest pupils, and private home schools for delicate or backward children of more advanced years. The Irish Catholic parent may send his children to the school of St. James, and the French Catholics have also a school of their own. Across the river the well known Bradford Academy and the Carleton School offer their advantages, while the famous schools of Andover and Exeter are reached by a few minute's ride. In neighboring towns the old Dummer Academy in Byfield, and the new Sanborn Academy in Kingston, each under most excellent management, invite to more quiet and peaceful halls of learning. The colleges of Harvard and Tufts, and the various institutions of Boston are an hour's ride away, and the railroad offers almost hourly facilities for reaching them.

But, up to the very entrance to the university, it is needless for any parent of Haverhill to seek training for his child elsewhere than in its public schools. What they may lack it is the purpose of the city to furnish, what they may do it is its purpose to do excellently, while in their breadth and extent of instruction it is its ambition to have them unexcelled, for it believes the public school to be the most powerful social factor in promoting its own material, moral, and intellectual well-being, and in magnifying and ennobling the gift of citizenship. *Quid munus Reipublicæ majus, meliusve, afferre possumus quam si juventutem docemus et bene erudimus!*

DEPARTMENT OF PUBLIC SCHOOLS.

NAME OF SCHOOL.	Number Teachers.	Enrollment of Pupils.	Rooms in Building.	Salaries of Teachers.	Valuation of Building.
High,	7	197	19	$7550	$75,000
Winter St. Grammar.	12	452	12	6850	30,000
Whittier "	4	156	4	2300	15,000
School St. "	8	282	7	4850	16,000
Currier "	12	474	12	6850	30,000
Portland St. Tra. School.	14	201	4	1880	15,000
Bowley,	5	205	7	3700	15,000
Wingate,	6	229	7	3250	16,000
Groveland St. Grammar.	4	157	5	2200	15,000
Locust St. Primary.	2	106	2	1000	6,000
Chestnut St. "	4	165	4	2100	15,000
Pond St. "	1	48	1	500	1,500
Tilton's Cor. "	1	34	1	500	1,500
Mill Vale Union,	1	19	1	500	800
Saunders' Hill Union.	1	13	1	500	800
Corliss Hill "	1	10	1	500	1,000
Rocks Village "	2	51	2	1000	2,000
Kenoza Avenue.	1	20	1	420	1,200
North Avenue.	1	25	1	500	1,000
North Main St.	1	30	1	500	1,800
Monument St.	2	72	2	1000	2,500
Broadway,	1	12	1	500	1,000
North Broadway.	1	17	1	420	1,000
Ayer's Village.	2	49	2	800	2,500
Lowell Avenue,	1	18	1	420	800
Eve'g School, Males,	7	146	6	600	
" " Females.	3	87	2	240	
" " Drawing,	2	65	2	300	
" " Book-keep.	1	33	1	120	

ORGANIZED ACTIVITY.

Haverhill's ecclesiastical history reads very much like that of so many of the older New England settlements, to whose inhabitants religious observances were meat and drink. Inspired to leave their native land and seek a lodgment in the wilderness by their inbred convictions in regard to the forms and methods of religion, it was inevitable, that there should be, to their minds, no distinction between religious and secular government, between taxes for police and taxes for preachers, between town and parish. Difficult as it may be for us to comprehend their intimate and inalienable association of the secular and the spiritual, to our minds and in our day so dissimilar, it was nevertheless for many years a fact, and a fact of great moment in the management of public affairs and in the growth of towns. In Haverhill, the town and the parish were identical for nearly a hundred and thirty years, town meetings and the services of the Sabbath being held in the same building, at once the townhouse and the parish meeting-house. In the town books and by the town clerk were preserved the records of such transactions as would now be considered as relating entirely to the various religious societies but which were then necessarily a part of the town's business.

In 1728 the town had become so large as to make it a matter of convenience for its inhabitants to divide it into two parishes, and later into more, so that those who lived in the remoter parts need not be obliged to go so far for worship. The parish meetings, having become, therefore, gatherings of a part of the people only, became also, perforce, distinct from the town meetings. Still, however, dwellers in the parish were, in the nature of things, by virtue of their residence, owners of the meeting-house, attendants upon its services, contributors to the support of preaching. At this time, and for forty years later, if any resident of the parish omitted to pay the parish rates, the parish collector was empowered to "take distress" on him and obtain the withheld rates by the sale of his seized goods. One John White, whose rates were gotten by this summary process, sued the parish to recover them, but lost his cause. The ancient rights of the parish being thus upheld, doubtless the parish officers were willing to become less strenuous in their exercise, and compromises were effected between the parish and its unwilling rate-payers until, a few years later, a special statute exempted from the payment of the parish rates such of the parishioners as presented to the authorities certificates of their membership of dissenting churches

FIRST PARISH CHURCH.

and of their payment of church rates therein. The relationship between town and parish continued so close that the town meetings were still held in the meeting-house of the first parish; and it was not until 1828 that the parish asserted its especial proprietorship by demanding payment from the town for the use of its building. Not until twenty years later did the town have a distinct assembling-place of its own. The difficulties in the way of calling parish and church synonymous were exemplified in Haverhill in the early part of this century by the disagreement between the Unitarian and Trinitarian wings of the Congregationalists, a familiar story in many New England towns. These occasioned a series of manœuverings for technical rights and possessions, and reached a climax at length in an open rupture between the two sects. Dissensions of the same general sort had arisen also in the West Parish, where the Universalists were more numerous, which were finally settled by mutual agreement as to which should be "the parish."

Already, however, in 1765, had occured the first break in the unanimity of religious worship in the town by the formation of a Baptist church, the first in the county, the evident declaration of what was to be a persistent rebellion against the traditional "standing order." It was not, however, until the next century that farther progress was made in the cultivation of a diversity of religious belief, but from that time on denominations arose and multiplied until now, in 1889, there are twenty-four church organizations, divided among eleven different denominations, — including Uni-

tarian and Trinitarian Congregationalists, Universalists,
Baptists, Freewill Baptists, Adventists, Methodists,
Episcopalians, Christians, Roman Catholics, and Spirit-
ualists. The Trinitarians have five churches scattered
over the city; the Baptists, five; the Unitarians, Uni-

CENTER CHURCH.

versalists, Episcopalians, Methodists, and Roman Catholics, two each; and the rest one. Some of these
edifices are remarkable for beauty and adaptedness.

The present church edifice of the First Parish was
built in 1847 to replace one that was destroyed by fire

on the first day of January of that year. When built it
was placed with the front to the south, but in 1884 the
structure was raised, enlarged, and turned to face the
east. At that time a large and commodious vestry was
constructed underneath, an addition made to the rear,
making room for the organ and choir, the old windows were replaced by rich and tasteful designs in
cathedral glass, the interior was frescoed in agreeable
colors, and the exterior painted in color similar to the
old red sandstone. The audience room has a seating
capacity of about 500. The vestry beneath, which has
assumed the name of Unity Hall, will accommodate
something more than 300. Both rooms are light and
airy, and furnish a convenient and desirable church
home for the men and women who worship there.
The church is situated on the corner of Main Street and
Crescent Place, immediately in front of the foot of
Summer Street. The present pastor began his labors
with the parish in October of 1881, and a good degree
of prosperity has attended the endeavors of the people.
Their hope is to make religion a practical application
of the principles of human brotherhood to the social
and business affairs of daily life, in the belief that integrity and sincere manliness are the foundations of all
success.

The Center church is located on Main Street directly opposite the City Hall. The edifice was completed
and dedicated on December 17, 1834. It originally
varied alike in appearance and arrangement from the
present structure. The entrance was adorned by two
massive pillars, "one on the right hand, the other on

the left," in imitation of those at the entrance of Solomon's temple at Jerusalem. The gable was ornamented by belfry and spire. In 1859 the old gallery

NORTH CHURCH.

was torn down, and the walls were finished in imitation of heavy stone work. The auditorium was enlarged in order to make room for additional pews. The tower and spire were built at this time. The church was re-dedicated January 27, 1860. In 1878 the interior of the church was again remodeled. The entire building was raised for the purpose of constructing in the basement a lecture-room, dining-room, ladies' parlor, kitchen, and library. A gallery was built across the western end of the auditorium, a new pulpit was furnished, and the walls and ceilings were appropriately frescoed at a cost of nine thousand dollars. The pastors of the church have been: The Rev. Joseph Whittlesey, installed Aug. 28, 1833; the Rev. Edward A. Lawrence, D. D., installed May 4, 1839; the Rev. Benjamin F. Hosford, installed May 21, 1845; the Rev. Theodore T. Munger, D. D., installed Jan. 6, 1864; the Rev. Charles M. Hyde, D. D., installed Nov. 15, 1870; the Rev. Henry E. Barnes, D. D., installed Nov. 23, 1876, and the present pastor, the Rev. Edwin C. Holman, installed Dec. 15, 1886.

The corner-stone of the North church, a substantial wooden edifice erected by a society which was an offshoot from that connected with the Center Church, was laid July 20, 1859, at the corner of Main and White streets, at the top of the hill which there rises from the river with a pretty steep ascent. It was dedicated Feb. 21, 1860. It is ninety feet long and sixty feet wide, containing one hundred and thirty-two pews, with a seating capacity of about seven hundred. It has in the basement a chapel, with parlor, kitchen, and smaller rooms; is finished with a tower, belfry, and spire, with a clock given in large part by residents of the neighborhood, and cost about $30,000. The first

pastor, the Rev. Raymond H. Seeley, D. D., remained with the church from his installation in 1860 until his lamented death in 1885, when the Rev. Nehemiah Boynton, who had been installed as associate pastor the previous year, assumed the pastorate. The latter resigned in 1888, and in 1889 the Rev. James W. Bixler was chosen his successor. The church has a membership of about five hundred, the Sunday-school of about six hundred; and the affairs of the society are in a very flourishing condition.

Trinity church was organized October 8, 1855, and the Rev. W. C. Brown was its first rector. The corner-stone of the present building, on White Street, was laid May 15, 1856, and the first service in the completed church was held on Christmas of that year. It was consecrated Jan. 7, 1857, by Bishop Eastburn. Upon Mr. Brown's resignation in 1858, the Rev. Charles H. Seymour became the rector. In 1865 an addition was made on the southerly side of the building, increasing the seating capacity to 500. Mr. Seymour resigned his position in 1868 and in July of that year the Rev. S. C. Thrall succeeded him. In 1869, by the exertions of the parish, with generous aid from citizens of the town and liberal donations from friends of the church abroad, a chime of bells was placed in the church tower, being at that time and for some years after the only chime of bells in Essex County. Dr. Thrall resigned in 1871 and was succeeded in 1872 by the Rev. Charles A. Rand. In 1880, the twenty-fifth anniversary of the formation of the parish was celebrated by a convention of the bishops and clergy of the diocese, and about a thousand dollars was expended in improving and beautifying the church. Mr. Rand's death in 1884, by the wrecking of the steamship on

FIRST BAPTIST CHURCH.

which he was journeying to Florida, ended a pastorate of twelve years and deprived the parish of a faithful and beloved teacher. In 1885, the present rector, the

Rev. David J. Ayers, assumed the pastoral charge. Many improvements have been made during the past few years. A rectory (the bequest of a former parishioner) has been added to the property of the church; a new organ has been placed in the chancel; an elaborate and costly font, a rood screen, a pulpit, a chancel rail, and many other beautiful gifts have added greatly to the beauty of the interior and to the convenience of worship.

After building three meeting-houses on Merrimack Street, the First Baptist Religious Society erected its present commodious and attractive home on Main Street in the year 1883. This church edifice may be classed with the largest of the state, the place covering about fourteen thousand square feet of land. The building is divided in plan into entrance porches and tower, auditorium, choir, and chapel. The tower is nineteen feet square at the base and rises one hundred and forty feet. The auditorium seats comfortably one thousand persons. With its rose windows, immense chandeliers, and large organ, this is one of the most beautiful houses of worship in New England. The whole property, including land and parsonage on Newcomb Street, has a value of $100,000. The twenty-three members who constituted this church in 1765 have increased to four hundred in 1889.

The first Universalist church edifice in Haverhill was built in 1825 and dedicated the thirtieth of November of that year. The dedicatory sermon was preached by the Rev. Hosea Ballou, and the prayer of dedication was made by the Rev. Thomas Whittemore.

FIRST UNIVERSALIST CHURCH.

The building was of brick, 40 feet in width by 55 feet in length. A steeple, heating aparatus, and bell were provided some years later. The present edifice was erected in 1855. It is of wood, 48 feet by 75 feet in dimension, and is located on the site of the old church on Summer Street, corner of Bartlett Avenue. It has a seating capacity of about five hundred. Some years ago a chapel was built beneath the church, and the audience room has been several times extensively renovated. A large and fine organ is now located on the right of the pulpit. The church has had fourteen pastors, the longest pastorate being that of the Rev. Calvin Damon, who, in two settlements, served the church for nineteen years. The present pastor is the Rev. J. C. Snow, D. D., who was called to the charge of the church in November, 1882, and entered upon his duties the following January. The congregation is of good size and embraces members of the prominent and influential families of the city.

In the fall of 1884 was laid the corner-stone of St. James' Roman Catholic church. This structure, built in the conventional Gothic style of architecture, is 175 feet long and 75 feet wide, with a seating capacity sufficient to accommodate 1400 people. Its steeple is 215 feet in height. The cost of this building when completed, which happy result it is expected will be reached within two years, will be $130,000. The whole edifice is pronounced by competent judges to be one of the finest pieces of architectural church work in New England. In addition to this there are connected with the parish a parochial residence, a convent,

ST. JAMES' ROMAN CATHOLIC CHURCH.

and a parochial school, all three possessing latest
modern improvements. The school building contains
16 class-rooms, with accommodations for 900 children.
The above furnish sufficient evidence, were such evidence needed, not only of the concord and harmony
existing between the pastor of St. James' Church, the
Rev. James O'Doherty, and his people, but also of the
sobriety of the people, their faithfulness to their employers, their steadiness at work, and their economy,
in being able to accomplish all this without any perceptible decrease in their savings.

There is something about the air and the soil of
these frontier settlements that breeds men, even if the
reluctant clearings are churlishly irresponsive to the
farmer's hand. Here in Haverhill, however, the land
was fertile both in crops and in men. At a time when
the clergy held the first place in the esteem of men,
ruled over their parishes with a sway more or less
autocratic, and walked out of church at the head of
their flocks, who waited in patient reverence until the
lordly cleric had passed by, Haverhill's divines ranked
with the best; and there have not been wanting those
since whose fame has not been limited to Haverhill or
its vicinity and among whom it would be invidious to
particularize. There are now in Haverhill about
thirty of each of the three learned professions.

In the colonial and provincial time the most important family, however, was that of the Saltonstalls,
whose intellect and capacity made them conspicuous
among their townsmen, and whose descendents, near or
remote, have well preserved the traditional reputation

WHITTIER'S BIRTHPLACE.

here and elsewhere. With the Revolution, however, came wider opportunities and greater necessities. The times made men. For this war, as for all the others, Haverhill furnished its full share of the rank and file and also men of the requisite stuff for higher duties. Its sons contributed to the roll of commissioned officers one colonel and four brigadier-generals of the Revolutionary army, the chief medical officer of the United States Army at the close of the Revolution, a brigadier-general of the war of 1812, and another general, "the most conspicuous soldier of Massachusetts" in the late war, himself grandson of one of Haverhill's most eminent men, Bailey Bartlett, for forty-one years high sheriff of Essex and four years a representative to the Congress of the United States. Later Haverhill sent to Washington another representative for four years, James H. Duncan; and the present United States senator from Kansas and president of the Senate is considered in Haverhill as properly one of its sons.

In other walks of life, however, the natives of Haverhill have sought and found distinction. Among them have been Daniel Appleton, founder of the well-known publishing house of D. Appleton and Company; Benjamin Greenleaf, excellent mathematician and author of a series of widely used mathematical text-books; Harriet Newell, one of the pioneers in the establishment of the missionary system in India and whose sad death at the age of nineteen hallows her memory. Haverhill's most distinguished scholar was undoubtedly Charles Short, at one time president of Kenyon College, Ohio, and at the time of his death professor of the

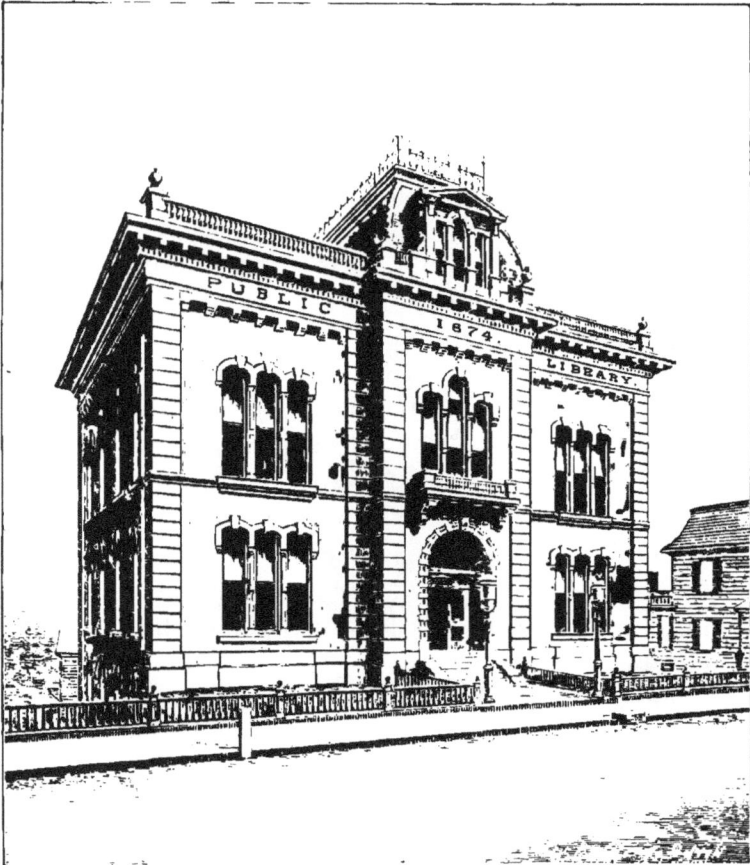

PUBLIC LIBRARY.

Latin language and literature at Columbia College, New York.

The most distinguished native of Haverhill is the poet John Greenleaf Whittier, who, born in a low-roofed farm-house, now two hundred years old, in the eastern parish of the town in 1807, spent here also his youth and early manhood. He worked on his father's farm, got the usual schooling of the country boy in the district where he lived, and, later, supplemented this scanty education by attendance upon the Haverhill Academy, where he himself afterwards taught. He edited one of the Haverhill papers for a time and then departed for a wider field of usefulness. To the home of his boyhood his heart has always turned. His poems breathe the air of Essex, and paint its landscape, its home life, its traditions. His birthplace, the Mecca to which the steps of reverent pilgrims turn each year, has been celebrated by himself in "Snow-Bound." It is in itself the simplest of natural scenes, not unfitting the simple nature of the man, a low and rude house standing near the road-side, where the stage-road to Amesbury is intersected by a cross-road. He describes the familiar scene as "the old farm-house nestling in its valley, hills stretching off to the south and green meadows to the east; the small stream which came noisily down its ravine, washing the old garden wall, and softly lapping on fallen stones and mossy roots of beeches and hemlocks; the tall sentinel poplars at the gateway; the oak forest, sweeping unbroken to the northern horizon; the grass-grown carriage-path, with its rude and crazy bridge."

RESIDENCE OF MR. J. H. WINCHELL.

The spot is even now secluded and peaceful, but far more lonely in his day, when it was scarce visited but by the weekly stage. Here, however, were formed the purity of soul, the unselfishness, the regard for principle, the love of freedom, and the carelessness of personal consequences that have marked his career. Here, also, he fellowshiped with the musk-rat and the squirrel, learned the sources of the brooks and their pathways to the river, drank in the "old wives' tales" of the neighborhood, and thus, in unconsciousness, wrapped the mantle of the poet around him. In his own words, —

> "I was rich in flowers and trees,
> Humming-birds and honey-bees;
> For my sport the squirrel played,
> Plied the snouted mole his spade;
> For my taste the blackberry cone
> Purpled over hedge and stone;
> Laughed the brook for my delight,
> Through the day and through the night,
> Whispering at the garden wall,
> Talked with me from fall to fall;
> Mine the sand-rimmed pickerel pond,
> Mine the walnut slopes beyond,
> Mine, on bending orchard trees,
> Apples of Hesperides!"

Whittier, though now a resident of Danvers or Amesbury, is beloved and revered in the city of his birth, where a club, formed in his honor, delights in remembering annually his birthday with some token of their regard.

ESTATE OF THE LATE C. D. HUNKING.

Of its public library Haverhill is, and may well be, proud. It is a perpetual monument to the liberality of its founder, the late E. J. M. Hale, who gave to the city the lot on which it stands, half the cost of the building, a legacy of a hundred thousand dollars, and other sums at various times, making a total of $174,-500. To this has recently been added a legacy of $15,000 from one of the trustees, lately deceased. In this institution were naturally absorbed the books of the Haverhill Library Association, which had hitherto endeavored to supply the public need for reading matter. Its elegant and commodious building was erected in an excellent location, in 1875, at a cost of $50,000, by a Haverhill builder, after the plans and under the supervision of a Haverhill architect. It is built of brick, having a frontage of seventy-two feet and a depth of fifty-five, with three stories, respectively twelve, sixteen, and twenty feet high. The ample basement is devoted to the reception and storage of books, etc.; on the eastern side, the upper stories are occupied by the circulating library and reading-room; while on the western side, the space of both stories is converted into two lofty halls, broken only by galleries, and used, one for the distribution of books and one for a reference library. The walls and ceilings are beautifully frescoed, and the rooms and halls are adorned with abundant pictures and busts, including one of the only two casts of Houdon's Washington after the original statue, a bust and an oil painting of Whittier, and many classic engravings.

RESIDENCE OF MR. C. W. ARNOLD.

The library contains 45,000 volumes and loaned in 1888, 58,132 books, or an average of 209 a day. The reading-room contains eighty-six newspapers, periodicals, and reviews, daily, weekly, monthly, and quarterly, and affords abundant opportunities to the student of contemporary literature. The books in the reference library have been selected with unusual care and are especially rich in the department of art. At the opening of the library, it was intrusted by the trustees to a gentleman of long experience in the public library of Boston, who has ever since continued in charge. The only condition imposed upon the city by the founder was that the city should meet the current expenses, and a large part of these are defrayed by the interest of a subsequent legacy from the founder, so that the annual cost to the city is but a trifling sum. It is, and is meant to be, of use to the student, the artist, the mechanic, and the casual reader; and it is an important factor in the social, educational, and literary growth of the city. It is not unlikely, that many of the social and literary clubs, for whose number Haverhill is celebrated, owe, if not their origin, the stimulus of their later growth and success, to the opportunities afforded by this library. Its facilities, for a city of this size, cannot be surpassed, or its value over-estimated.

Haverhill is a city in which the average man appears to good advantage and in which the extremes of riches and poverty do not abound. Poverty and riches exist, but not in the marked antithesis that obtains in some communities. Being thus a place in

RESIDENCE OF MR. S. P. GARDNER.

which there are many "well-to-do" but few of the
very rich or the very poor, it is essentially a demo-
cratic city, where equality obtains without the need of
offensive self-assertion. Many of its wealthy men have
themselves worked, at the bench or elsewhere, and
attained riches and position by their own exertions,
and are thus naturally in touch with those who are
likely, later on, to come from the same bench to take
their places. It follows, therefore, that there are comparatively few residences conspicuous among their
fellows for lavish architecture or luxurious adornment,
although some of the less pretentious are noteworthy
for the evidence of an artistic sense and a trained taste
in their furnishing. The stranger within Haverhill's
gates is, however, always taken to drive through
" Birchbrow," the estate of Mr. Thomas Sanders, the
present president of the Board of Trade, and to " Win-
nikenni Hall," until recently owned by the late Dr.
James R. Nichols, who came to this city on foot, a
farmer's boy, to seek his fortune, and now the property
of Mr. William G. Webb of Salem. Each of these
charming homes rises from one of Haverhill's abundant hills to overlook a lake, and each bears witness,
in location and structure, to the good taste of the
builder. Open to the people of the city by the liberality of their owners, their grounds are practically public parks, their four or five miles of roads affording a
various landscape of hill, valley, and lake. Among residences less removed from the more compact part of the
city, but illustrative of comfortable dwellings and the
home-building spirit of the people, are those of Mr.

RESIDENCE OF MR. J. M. DAVIS.

James H. Winchell, at the corner of Pleasant and Peeker Streets, in the older part of the city; of the Hunking estate, on the main avenue leading northward from the bridge, and of Mr. C. W. Arnold, some half mile beyond; of Mr. S. Porter Gardner, recently erected on a very sightly elevation on the " Highlands;" of Mr. Jared M. Davis, in the thriving and finely located village of " Riverside."

Haverhill, though not, as was said above, one of the places " where wealth accumulates and men decay," has yet some poor and unfortunate, and, having the occasion, has also the willingness and the capacity to provide for them. The Female Benevolent Society, which came into being soon after the war of 1812, has ever since been active in measures of relief for the needy and is still cordially supported as one of the institutions of the city as well as for the good its more than three hundred members accomplish.

There was begun in 1858 a society since merged in the Old Ladies' Home Association, which was designed to provide for such women as might need it a home for their declining years. A commodious building, easy of access, was built for the purpose in 1876 at a cost of ten thousand dollars, and has since afforded a comfortable retreat for many worthy women. The society has funds to the amount of nearly thirty thousand dollars, exclusive of the Home.

The Children's Aid Society was the outgrowth of a disposition among some of the benevolent women of the city to provide destitute children with the care and comfort of home, whose plans were brought to a head

OLD LADIES' HOME.

by its formation in 1865. It was not, however, until 1871 that a building was obtained, to be used for the purpose; and this was replaced in 1884 with a handsome brick building at a cost of twenty thousand dollars. The society, which was formed and is managed by women, now holds to its credit a fund of over fourteen thousand five hundred dollars, and real estate, exclusive of the Home, valued at ten thousand dollars. It has a hundred life members and over three hundred annual members, and maintains in its comfortable quarters over thirty-five children each year.

The City Hospital owes its origin to the late E. J. M. Hale, who left to trustees a fund of fifty thousand dollars and a site for a hospital. The trustees were organized in 1882, but no active steps were taken until five years later, when another site was presented to the trustees by Mr. James H. Carleton. The buildings on it were at once remodeled for hospital purposes and formally opened in the last week of 1887. The hospital is managed by seven trustees, of whom the mayor of the city is *ex officio* chairman, has an attending staff of six surgeons, and a consulting staff of five, and affords accommodations for thirty patients. Situated upon elevated land about a mile from the city, commanding a view of two lakes, it is admirably adapted to afford to its inmates cheerful surroundings and abundance of fresh air. The trustees still own the original site given them by Mr. Hale and derive the means for the maintenance of the hospital in part from the proceeds of their invested funds and in part from the contributions of the charitable, who take a deep interest in the hos-

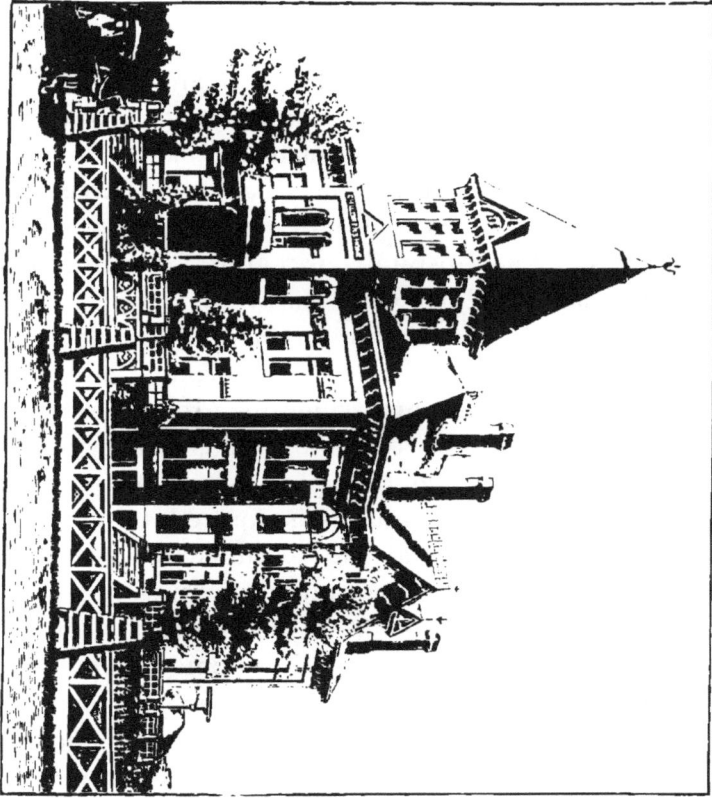

CHILDREN'S HOME.

pital as doing a needed and practical work. Within
ten days of its opening its accommodations were taxed
to the utmost by a terrible disaster upon the Boston and
Maine railway, just across the river from Haverhill, by
which thirty persons were injured and fourteen lost
their lives. The hospital has cared for a little over a
hundred patients during the past year.

Haverhill is remarkable for the number and variety
of its clubs, — clubs of men and of women and of both,
clubs for social, literary, scientific, religious, medical,
legal, and culinary purposes. If it is hoped to further
a "cause," to improve the mind, or to pass an occasional pleasant hour, a club is formed to do it. The
whole network of social life is interwoven with clubs.
Most noteworthy, perhaps, among them is the Monday
Evening Club, an association of gentlemen formed in
1860, which has included in its membership many of
Haverhill's foremost citizens and which has lost to-day
none of its prestige. It has set a pattern which other
cities have copied with advantage in the formation of
similar associations, and there is also a second in
Haverhill, the Fortnightly Club, founded after its
fashion. These and like organizations serve as the
useful stimulus to study and culture that every manufacturing town is apt to need, and keep lit the flame of
literature, early kindled here. During the siege of
Boston, some of its well-born families, driven thence,
found a warm welcome and a happy refuge here; and
some staid on after the siege was raised, thus increasing the number of the cultured. In the early days of
the Revolution, also, a part of the library of Harvard

"CITY HOSPITAL."

College was brought here for safe keeping, and it was even proposed to move the college here.

Among the organizations of which sociability is the more distinctive feature is the Kenoza club, an association of gentlemen, who occupy a picturesque spot on the shore of Lake Kenoza. There is a grove, a house, with conveniences for cooking; and the place affords a pleasant retreat, not too far removed, from the noise and dust of the city. It is a favorite resort of pleasure parties and is likely to become more so, as the members of the club have in view various projects for increasing the attractiveness of the grounds and extending the facilities for boating and fishing.

CITY HALL REBUILT IN 1889.

SHOES AND SHOEMAKING.

Years ago, about the beginning of the nineteenth century, when Haverhill laid off its swaddling-clothes as a village and assumed the dignity of a town, it was noteworthy as a market-place. On the bright summer days the principal street of the place was filled with the wagons of the farmers who came in from New Hampshire, and even from the far off hills of the Green Mountain state, to exchange their produce for other necessities of life; and it was from this that the shoe business, as a business, had its origin. Throughout all the little hamlets that are scattered over the granite hills of New Hampshire could in those days, and can even now, be found little shops of one room each, in which the sturdy farmers eked out the existence which they with difficulty maintained upon their scanty farms. Throughout the winter months these workmen toiled over the lap-stone, making the shoes which, with the advent of spring, found a ready market

in the town.   The transition from this state of affairs to
the concentration of the business in the town itself was
a natural if somewhat slow one.

The shoes thus made were no sooner seen than
appreciated.   They were well and honestly made, of
good material, and for durability and looks could not
be surpassed in any section of the country.   The
demand soon exceeded the supply, and, consequently,
some effort must be made to increase the production.
Moreover, the younger members of these artizans'
families were ambitious.   They longed for some wider
field of action and were not satisfied to tread the paths
their fathers trod, to live confined within the narrow
circumference of their native village, while, naturally
enough, there was not room for them within the walls
of the old homestead.   The "town" offered them
greater possibilities, and it was to the town that their
steps naturally turned.   The result was inevitable.
Haverhill shoes were in demand.   By combining their
efforts, working constantly and with system, with a
supply of material afforded by increased capital, two
men could accomplish, in the town, what four men
could not do on the isolated country farms; and thus it
was that the first shoe manufactory was established
within the limits of Haverhill itself.

But limited capital, comparatively speaking, was required.   In those days the jobbers sought the manufacturers and every Haverhill establishment was sure that
its products would at once find ready sale.   For years
there was "nothing like leather," and, although competition existed, although Haverhill was not alone in

finding out the advantages of the trade, yet it can be
said in all sincerity that the quality of the work done
here was of a far higher order than that done elsewhere. The same characteristics which marked the
shoes made in spare time devoted to their manufacture
by the farmer shoemakers existed in the goods turned
out from the manufactories. They were hand-made, reliable, stylish, "fine" goods. Gradually the town grew;
the immigration to it from the surrounding country
increased; new factories were opened; men with no
other capital than their sturdy arms, inbred knowledge
of their trade, and courage started out in business for
themselves, made by their own labor their samples,
and, when they found a sale for their goods, hired
assistance as it was needed, increasing their facilities
as the business grew. Then some happy genius
bethought himself that more work could be done if it
was sub-divided into its natural divisions, if, instead of
one man making the whole shoe in all its details — cutting, sewing, lasting, etc. — from start to completion,
one should devote himself wholly to cutting up the
stock, another to fitting it, and so on. Thus originated
the so called "team" work, — five, six, or more men who
banded themselves together, and, either as employers
or on the co-operative plan, were able to do much more
work, with far greater comfort and ease to themselves.
Thus the industry made a slow natural growth, ever
increasing, but with far from giant strides, until the
great civil war broke out. The impetus given to the
trade and in fact to all other trades by this "blessing in
disguise," for, despite the horrors, sorrow, and indi-

vidual misery caused by it, it is useless to deny that the
full resources and possibilities of our country were
brought out and marvelously expanded by the War of
the Rebellion, is too well known to demand any extended notice in an article of this character.

The impulse thus given has, however, completely
revolutionized the shoe business. The introduction of
machinery has enabled the production to be enormous-
ly increased, while at the same time it has lowered its
cost. It has done away entirely with the old order of
things, "team" work no longer has an existence, while
a successful manufacturer of fifty or even twenty-five
years ago would find himself entirely at a loss to comprehend or carry out the various ramifications which are
now the ordinary details of the trade. Some faint idea
of how the business has grown may be gleaned from
the statement, that in 1832 there were twenty-eight
firms engaged in the manufacture of shoes in Haverhill; in 1837, forty-two; while at the present time
there are fully two hundred firms, giving employment
to fifteen thousand operatives, distributing annually
more than $2,000,000 in wages, and shipping each
year over two hundred thousand cases, the shipment
the past year reaching the enormous total of 250,338
cases.

Through all the time, amid adversity and prosperity, in good years and poor years, the city has always
maintained its pristine reputation for turning out fine
goods; and to-day, unlike all other shoe manufacturing
places in the state, Haverhill prides itself, not so much
on turning out more shoes per annum than any other

city in the world, but on the fact that the bulk of the
trade is in "fine" goods, whether hand or machine
made. Almost every variety of leather foot-wear
which the ingenuity of man can devise is manufactured
here, including men's, women's, and children's boots,
shoes, and slippers; and some of these are of the so
called "cheap" goods. The latter is comparatively
a new departure, for, as before stated, Haverhill made
its reputation distinctively on fine hand-sewed goods, and
it was the skill and artistic taste displayed by the leading exponents of that art of which St. Crispin is the
patron saint that caused the steady and rapid increase
of business and the consequent growth and prosperity
of the city. Every device that would add beauty to
the appearance, comfort in the wearing, and that inex-
pressible attribute that the French call *chic* has been
studied out and adopted by the local manufacturers.

The ingenuity of that most ingenious race, the New
England Yankee, has been taxed to the utmost under
the fierce heat of competition, and the artistic beauty
of many of the Haverhill shoes is without equal in their
line, while the talent and skill displayed by the authors
of the various conceits could hardly have failed to have
given their possessors fortune and reputation, no matter in what line of life's work exerted. This, with the
fact that the shoemaker of the type of which Haverhill
boasts is a practical mechanic, being born and bred to
the business, is what, to-day, makes New England's
shoes find a ready market all over the country and
prevents the industry from being transplanted, to any
great extent, to other sections of the country. For the

most part, the men now at the head of the most prosperous and active, and, consequently, largest manufactories of the city were once day laborers at the bench,
and their acquaintance with the various departments of
shoemaking is a thoroughly practical and often a very
useful one, many a Haverhill shoe manufacturer step-
ping in, in an emergency, and taking hold of some
abandoned department of the work.

As the character of the business has changed so has
its needs, increased accommodations have been demanded and supplied, and, as a result, the visitor to the
once little market-town is confronted with acre after
acre of handsome brick blocks of five and six stories
fitted up expressly to accommodate the business of making
shoes.　To a casual visitor this is the first thing which
attracts attention, and yet, should he visit the suburbs,
he would find other manufactories of an even higher
grade as far as convenience and adaptability to the
purposes for which they are used is concerned.　And
this brings up a feature which Haverhill claims, and
justly, will ensure the continued growth and extension
of the business.　In all quarters of the city, convenient
of access, are desirable locations on which can be
erected factories which are just far enough from retail
business life to be cheap in rent, isolated enough to
bring insurance to a minimum, and commodious
enough to furnish all possible accommodations, while
at the same time capitalists stand ready and willing to
build such buildings as are wanted.

The fact is, the tendency of the times is toward
concentration, manufacturers are realizing more and

more the almost vital necessity of having all their work done under their own supervision and in their own buildings; and it is this that promises, most of all, to tend to the future growth of the shoe business in Haverhill and of the city itself, since it promises to weaken the force of two competitors that have confronted it in the past, — the country factory and the country workman. At the present time, the ten acres of closely crowded brick blocks, the isolated factories scattered through the suburbs, represent but a part of the shoe industry of the city; and any sketch of that would be incomplete did it fail to mention the fact that a large proportion of the shoes sold by Haverhill manufacturers were made outside the city limits. The same casual observer, if he continued to inspect the city, could not fail to notice, drawn up before a factory door, a huge express wagon, attached to which are four horses, and loaded down with shoe-boxes. The team is that of a "freighter," so called, and it is receiving boxes of unfinished shoes, to be carried to some country town in New Hampshire to be made up in just the same manner and in, perhaps, just the same kind of shops and by the same class of workmen as were mentioned in the earlier pages of this article, in recounting the origin of the shoe business in Haverhill. There are some fifteen or twenty of these freighters, making trips to the city daily or several times a week, covering distances of ten, twenty, forty, and even sixty miles, and in every little hamlet through which they pass leaving materials to be made up into shoes, on their homeward trip, and taking away the finished shoes on the journey to Haverhill. This business, while it has long outlasted its fellow, the old-fashioned stage line, is rapidly losing its comparative importance. " The mountain will not go to Mahomet, so Mahomet must go to the mountain;" and the relative decrease in the freighting of shoes means, that the workmen are coming to the city, and that, under the present competition, the cheapest and best work is done directly under the eye of the manufacturer.

There have been in Haverhill, and are even now, occasional desertions from the ranks of the manufacturers by those whose energy and ambition hope to find elsewhere better fields for the display of these qualities, who look for less embarrassments, larger opportunities, more compliant workmen, and who expect better returns for their invested capital, elsewhere than in Haverhill. But they look in vain. The endeavor is as futile as to stem the natural current of population city-ward. The same tendency which settles half of our inhabitants in cities has its influence in determining the centers of manufacture, and they who oppose it strive in vain. The early shoemaker, as has been said, was the owner of a farm, the possessor of land, who supplemented the rewards of this calling by the proceeds of another and whose attachment to and ownership of his home determined his permanent residence there. With the great increase in demand and production of goods, and the necessary multiplication of workmen, arose a proportionate number of shoemakers who had no homes of their own until in later years their accumulated wages supplied them. These naturally drifted to the city, where work was more likely to be plenty and permanent, where boarding-places abounded, and where the opportunities of a city offered what would be to them alluring advantages. Thus, in the nature of things, the city throve at the expense of the country.

This is not, however, the only or the main reason for Haverhill's growth. It is a familiar maxim that "nothing succeeds like success," and, it having once been known and understood that Haverhill was a center for the manufacture of "fine" goods, the best workmen, when in search of employment, turned their steps hither, expecting to find work and wages proportioned to their skill. The custom, once formed, always obtained. And thus, as the great corporations of Manchester, Lowell, Fall River, can make cottons and woolens to the best advantage; as the carriage-builder of Amesbury and Merrimac can make the same vehicle cheaper than his competitors in places where carriage-making is not the main industry; as, in general, the best results are obtained at the lowest cost where skilled labor naturally congregates; so in Haverhill the maker of shoes can turn stock into manufactured goods better and cheaper here because the skilled workmen are drawn to his factory by a natural law. The concentration of skilled labor at certain points, in obedience to forces that cannot always be defined, but which can never be successfully opposed, has made possible the origin and growth of the industrial centers of New England.

The business of shoemaking once well established here, the dictates of convenience, economy, and good business management alike suggest to the manufacturer the advantage of pursuing it in Haverhill. If a man falls sick, another skilled workman, not a mere stop-gap, is ready to take his place to carry on his familiar work. Does the machinery break down, other manufacturers are ready to lend, or the broken pieces can be supplied from neighboring stores at a moment's notice. If a shortage arises in one or another of the various odds and ends that enter into the making of a shoe, all of them can be had at once from the stock dealer who finds his opportunity in the aggregation of manufacturers and the consequent demand for material. Nor is the gain in convenience alone in the proximity of the stock dealer. The manufacturer, being on the spot and forced to buy only according to his present need, can take advantage of the market, while he who lives at a distance must carry a stock much of the time needlessly large or run the risk of coming short at an inconvenient season.

The banks, too, preferring to lend money to the city manufacturer, favor him by lending to him at lower rates than to his competitor in the country. The railway is at his door. The capital of the state, where the buyers of shoes gather from all over the country, is but an hour's ride; and the intercourse of maker and buyer is therefore easily had, common, frequent. When, therefore, one stops to consider the distance of the country manufacturer from the abundance of skilled workmen and from his source of supplies, the wear and tear

involved in the carriage of goods, the extra travel and
inconvenience occasioned by the distance from the
centers of trade, and the balance of the interest account
against him, it is small wonder, that, while the vast
increase in manufacture has been evident both in city
and country, the far greater proportionate increase has
been, and is likely to be, in the former, and that, in the
course of years, it is not impossible that the shoe
freighter may find his occupation gone.

Nor is it strange that a city so well located as
Haverhill should invite and retain capital, to be invested in manufacturing. Land in almost all parts of
the city can be had for resident purposes at reasonable
prices; two co-operative banks are ready and willing to
assist every workman to become his own landlord;
while the cost of living is quite as small as in the
smaller towns. And, moreover, it is from the workingmen themselves that the ranks of the manufacturers are
recruited. The hills of Haverhill are dotted with the
cottages of shrewd, intelligent, hard-working mechanics, who understand their business, who are ambitious,
and who realize that the world is their oyster and that
it can be opened by them to their future advantage, if
only they persevere. It is not alone its rich men, its
well-to-do manufacturers, that make Haverhill's shoes
hold their own, and more, in the market, nor is the
growth of the city because of them, but it is because
the majority of its skilled workmen have a personal
interest in its welfare, and are likely to become tax-
payers, and so the best of citizens. Since, therefore,
the introduction of machinery has entirely changed the

character of the work and the methods of conducting
it; since the condition of the working man is bene-
fited; since economy of time and economy of material are both conserved, — it is not surprising that
the city prospers, and that the progress of events has
shown most conclusively, that, ere the beginning of the
twentieth century, — Haverhill will boast of factories
three and four times as large as any now built,
factories where every single constituent which goes to
make up foot-wear will be kept and where every iota
of the manufacture will be carried on directly under
the management of the manufacturer and his agents.
The growth of the shoe business from a retail to a
wholesale one has been slow, but it has been certain,
and the results are now beginning to be seen.

In addition to the manufacturing proper, there are a
hundred, nay two hundred, establishments within the
limits of the city that are connected directly with the
shoe business, outgrowths of it, and at the same time
strong props and stays to its perpetuity and growth,
since in them are sold the thousand and one parts that
are used in making up the simple-looking, but in
reality complicated, foot-covering, the machinery, tools,
and so forth used in its construction. There are
dealers in patterns, trimmings, dies, lasts, cut soles,
leather of all descriptions, rands, heels, tops, stiffen-
ings, wooden and paper boxes, leather-board, paints,
varnishings, and hundreds of other minor essentials.
In this connection it can be said, that Haverhill boasts
the largest sole-leather establishment in the United
States; that three firms employing in the aggregate

four hundred hands are engaged in making the paper
boxes and cartons in which the shoes are packed; that
there are three firms turning out the wooden "cases"
in which they are shipped; that one firm has a large
and paying business in making and planing the boards
used by the cutters in cutting up the skins for the boots
and shoes; that there are two factories busily employed
in the manufacture of nails for shoes. In fact, Haverhill is one vast shoe manufactory, its very life, existence, and prosperity dependent on the trade which has
made it what it is and on which it bases to a large
extent its hopes in the future.

The following statistics, which have been most
carefully compiled and are believed to be as nearly
correct as possible, will give, far better than mere
words can, an idea of the enormous amount of raw
material used, the manufactured product turned out,
and the hundred and one details which all unite in this
most interesting industry. A careful study of them
will well repay the reader, and, after reading the solid
mass of figures, he can easily see how deeply rooted is
the industry and what a vital part it plays in the
economic life of the city.

CLASSIFICATION OF STOCK USED YEARLY
IN THE BOOT AND SHOE INDUSTRY
IN THE CITY OF HAVERHILL.

ARTICLES.	BASIS.	QUANTITY.
Bags, Paper Packing,		1,800
Beaver.	Yards,	360
Board, Leather and Straw,	Pounds,	12,420
Board, Leather and Straw,	Pairs,	1,756,398
Boxes, Paper and Wooden,		3,772,572
Box Toes,	Pairs,	264,000
Bows,	Dozen Pairs,	3,360
Brocade,	Yards,	600
Buckram,	"	2,964
Cassimere and Felt,	"	2,052
Cement, Glue, Paste, Etc.,	Gallons,	34,349
Cloth, Cotton,	Yards,	866,467
Cloth, Emery,	Reams,	2
Cloth, Enamel,	Cases,	780
Cloth, Patent,	Yards,	15,600
Cotton Thread and Silk,	Pounds,	28,827
Cord, Clark's,	Balls,	25
Counters,	Pairs,	930,000
Cutting (contract work),	Cases,	360
Embroideries, Velvet,	Dozen Pairs,	326
Eyelets,		28,282,000
Findings, (costing)	Dollars,	1,467,877
Galloons, Cotton and Silk,	Gross,	2,880
Goring,	Yards,	14,400
Gum Tragacanth,	Pounds,	14
Heels,	Cases,	5,358
Heels,	Pairs,	777,960
Heels,	Sets,	180
Lacings,	Gross,	5,615
Linings,	Yards,	5,880

ARTICLES.	BASIS.	QUANTITY.
Linings,	Pounds,	360
Leather, Alligator,	Feet,	66.720
Leather, Buff,	"	1.432.310
Leather, Calf.	"	199,680
Leather, Chamois.	"	6,000
Leather, Dongola,		198.240
Leather, English.		3.600
Leather, Goat,	"	1,115,530
Leather, Grain,	"	734,950
Leather, Kid.	"	4.824,000
Leather, Mole Skin,	"	1.800
Leather, Mule,		1,800
Leather, Patent,	"	107.272
Leather, Sheep.		1.447.622
Leather. total,	Feet.	10.145.388
Leather. Alligator.	Dozens.	282
Leather, Hair Calf.	"	234
Leather, Calf Skins.		1,800
Leather, Dongola.		738
Leather, Kid (Bronze and French).	"	685
Leather, Patent.	"	2,400
Leather (Glove and Russet).	"	58.942
Leather. total by	Dozens.	65.080
Leather, Soles, Inner,	Pairs.	3.926.436
Leather, Soles. Outer.	"	14.031.454
Leather, Rands,	"	365.600
Leather, total,	Pairs,	18.323.492
Leather, Bellies,	Pounds.	192.000
Leather, Calf, French.	"	960
Leather, Calf. Plain.	"	102.501
Leather, Calf, Wax,	"	24,960
Leather, Kip,	"	12,360
Leather, Rands, Roundings, and Skivers,	"	3,769,819

ARTICLES.	BASIS.	QUANTITY.
Leather, Rounding,	Pounds,	115,200
Leather, Rough,	"	542,520
Leather, Scraps,	"	88,800
Leather, Sole,	"	25,020
Leather, Split,	"	254,894
Leather, Trimmings,	"	360,000
Leather, Uppers,	"	180,000
Leather, total,	Pounds,	5,639,034
Leather, Buff,	Sides,	1,200
Leather, Sole,	"	124,397
Leather, total,	Sides,	125,597
Needles,	Boxes,	482
Patterns, Velvet Slippers,	Dozen,	13,710
Plush,	Yards,	420
Satine,	"	60
Satin and Serge,	"	154,788
Shanks, Steel,		601,272
Stiffenings,	Pairs,	3,104,610
Stiffenings,	Sets,	250
Velvet and Velveteen,	Yards,	36,780
Webbing, Elastic,	"	1,200

CLASSIFICATION OF

GOODS MADE AND WORK DONE

IN A YEAR IN THE BOOT AND SHOE TRADE OF THE CITY OF HAVERHILL.

ARTICLES.	BASIS.	QUANTITY.
Bead Work,	Pieces,	6,000
Binding,	Barrels,	300
Boots, Boys' Calf Balmoral,	Pairs,	3,600
Boots, Boys' Calf Button,	"	3,600
Boots, Boys' Congress,	"	72
Boots, Boys', Miscellaneous,	"	17,952
Boots, Boys', total,		19,224
Boots, Children's Buff Polish,	"	360
Boots, Children's Goat Button,	"	5,040
Boots, Children's Goat Polish,	"	1,440
Boots, Children's Grain Button,	"	2,160
Boots, Children's Grain Polish,	"	360
Boots, Children's Kid Button,	"	5,400
Boots, Children's Kid Button and Lace,	"	13,200
Boots, Children's Kid Low Cut,	"	3,600
Boots, Children's Miscellaneous,	"	72,576
Boots, Children's Turned,	"	4,320
Boots, Children's Woolen and Felt,	"	3,600
Boots, Children's, total,	"	112,056
Boots, Men's Buff Balmoral,	"	57,744
Boots, Men's Buff Congress,	"	21,744
Boots, Men's Buff Polish Button,	"	28,944
Boots, Men's Calf Balmoral,	"	87,888
Boots, Men's Calf Brogans,	"	360
Boots, Men's Calf Congress,	"	13,584
Boots, Men's Calf Crimped,	"	360

ARTICLES.	BASIS.	QUANTITY.
Boots, Men's Calf Don Pedro,	Pairs,	173
Boots, Men's Calf Glove,	"	2,880
Boots, Men's Calf Opera,	"	1,440
Boots, Men's Calf Polish,	"	8,640
Boots, Men's Calf Polish Button,	"	84,000
Boots, Men's Calf Hand-sewed	"	1,200
Boots, Men's Custom Made,	"	800
Boots, Men's Dongola Button,	"	1,167
Boots, Men's Dongola Congress,	"	6,900
Boots, Men's Dongola Congress Foxed,	"	288
Boots, Men's Dongola Patent Dressed,	"	720
Boots, Men's Kip Brogans,	"	8,640
Boots, Men's Miscellaneous,	"	632,947
Boots, Men's Kip Hand-sewed,	"	13,500
Boots, Men's Machine Made,	"	89,728
Boots, Men's Patent Foxed Congress,	"	5,472
Boots, Men's Split Balmoral,	"	7,200
Boots, Men's Split Balmoral and Congress,	"	1,152
Boots, Men's Split Button,	"	3,600
Boots, Men's, total,	"	1,114,740
Boots, Misses' Buff Polish,	"	360
Boots, Misses' Calf Congress,	"	36,000
Boots, Misses' Dongola Button,	"	1,800
Boots, Misses' Goat Button,	"	10,152
Boots, Misses' Goat, Grain and Kid Button,	"	72,000
Boots, Misses' Goat and Kid Button,	"	2,700
Boots, Misses' Goat Polish,	"	1,656
Boots, Misses' Grain Button,	"	1,800
Boots, Misses' Grain and Kid Button,	"	7,632
Boots, Misses' Grain and Polish,	"	7,200
Boots, Misses' Kid Button,	"	900
Boots, Misses' Kid, Goat, and Polish,	"	1,800
Boots, Misses' Kip,	"	360
Boots, Misses' Miscellaneous,	"	78,600
Boots, Misses' Kip Hand-sewed,	"	180,000

ARTICLES.	BASIS.	QUANTITY.
Boots, Misses' Kip Button and Low Cut,	Pairs,	7.200
Boots, Misses', total.	"	410,160
Boots, Women's Buff and Calf.	"	58,800
Boots, Women's Buff India,	"	18,000
Boots, Women's Buff Polish,	"	150,408
Boots, Women's Buskins,	"	19,800
Boots, Women's Calf Balmoral,	"	12,960
Boots, Women's Calf Glove.	"	108,720
Boots, Women's Calf Glove Button,		50,400
Boots, Women's Calf Glove Low Cut Button,	"	5,400
Boots, Women's Calf Glove Congress.	"	1,080
Boots, Women's Calf Glove, Kid Foxed,	"	6,840
Boots, Women's Calf Glove Polish,	"	3,960
Boots, Women's Calf Polish,	"	32,400
Boots, Women's Dongola Button.		12,240
Boots. Women's Dongola Button Foxed,	"	2,520
Boots, Women's Dongola Polish,	"	2,160
Boots, Women's Dongola Polish Foxed.	"	1,080
Boots. Women's Goat.	"	161,832
Boots, Women's Goat Button,	"	93,528
Boots, Women's Goat Imitation Button.	"	2,160
Boots, Women's Goat Pebble,	"	100,800
Boots, Women's Goat Imitation Pebble.	"	4,320
Boots, Women's Goat Polish.	"	9,576
Boots, Women's Grain,		22,680
Boots, Women's Grain Button.	"	38,880
Boots, Women's Glove Grain,	"	30,960
Boots, Women's Glove Grain Polish.	"	103,320
Boots, Women's Kid.	"	43,632
Boots, Women's Kid Button.	"	564,312
Boots, Women's Kid Foxed Buskins.	"	720
Boots, Women's Kid Foxed, Glove Top.	"	4,320
Boots, Women's Kid, French and American.	"	36,000
Boots, Women's Kid, Glove Top.	"	2,160
Boots, Women's Kid India,	"	22,320

ARTICLES.	BASIS.	QUANTITY.
Boots, Women's Kid Laced,	Pairs,	14,400
Boots, Women's Kid Polish,	"	37,488
Boots, Women's Glove Kid, Calf Foxed,	"	1,440
Boots, Women's Glove Kid, Foxed Polish,	"	1,800
Boots, Women's Glove Polish Split,	"	46,080
Boots, Women's Spanish and Polish,	"	25,200
Boots, Women's Polish, Glove Top,	"	3,600
Boots, Women's Miscellaneous,	"	133,320
Boots, Women's Serge,	"	1,080
Boots, Women's Serge Balmoral,	"	216
Boots, Women's Serge Button, Foxed,	"	14,400
Boots, Women's Serge and Congress,	"	720
Boots, Women's Serge and Polish,	"	3,600
Boots, Women's, total,	"	2,032,122
Boots, Youths' Balmoral,	"	4,320
Boots, Youths' Calf Balmoral,	"	2,160
Boots, Youths' Calf Button,	"	6,080
Boots, Youths' Calf Congress,	"	360
Boots, Youths' Miscellaneous,	"	4,752
Boots, Youths', total,	"	17,672
Bows,	Dozen,	30,000
Button Holes,	——	1,096,168
Counters,	Pairs,	300,000
Doublers,	"	56,260
Fitting Boots and Slippers,	"	762,000
Heels,	"	1,082,000
Heels,	Barrels,	8,760
Heels, Pasted,	Pairs,	288,000
Heeling Boots, Shoes and Slippers,	Dollars,	90,642
Heel Stock,	"	480
Patterns, Embroidered Slippers,		36,000
Shoes, Boys',	Pairs,	4,752
Shoes, Children's Goat Oxford,	"	360
Shoes, Children's Kid Button Newport,	"	720

ARTICLES.	BASIS.	QUANTITY.
Shoes, Children's Kid Tie Newport,	Pairs,	360
Shoes, Children's Kid and Goat Ties,	"	14,400
Shoes, Children's Red Oxford,	"	1,080
Shoes, Children's Miscellaneous,	"	24,000
Shoes, Children's Woolen and Felt,	"	3,600
Shoes, Children's, total,	"	44,520
Shoes, Men's Calf Oxford Button,	"	5,184
Shoes, Men's Calf Oxford Ties,	"	1,440
Shoes, Men's Calf Strap,	"	2,880
Shoes, Men's Calf Sailor,	"	187
Shoes, Men's Dongola,	"	8,400
Shoes, Men's Dongola Ties,	"	5,760
Shoes, Men's Goat Ties,	"	864
Shoes, Men's Goat Low Shoes,	"	432
Shoes, Men's Goat Pumps,	"	720
Shoes, Men's Grain Low Shoes.	"	864
Shoes, Men's Grain Harvard Ties,	"	360
Shoes, Men's Kid Oxford Ties,	"	432
Shoes, Men's Kid Pumps,	"	1,152
Shoes, Men's Patent Leather Oxford Button,	"	360
Shoes, Men's Patent Leather Oxford Ties,	"	11,920
Shoes, Men's Patent Leather Pumps.	"	16,200
Shoes, Men's Velvet Oxford Ties,	"	3,200
Shoes, Men's Velvet Pumps,	"	1,800
Shoes, Men's Velvet Ties,	"	720
Shoes, Men's Miscellaneous,	"	709,245
Shoes, Men's Miscellaneous Custom Made,	"	800
Shoes, Men's Miscellaneous Hand-sewed,	"	13,500
Shoes, Men's, total.	"	787,880
Shoes, Misses' Buff Low,	"	1,008
Shoes, Misses' Alligator Oxford Ties Im't,	"	1,800
Shoes, Misses' Goat Harvard Ties,	"	864
Shoes, Misses' Goat Newport Button,	"	720
Shoes, Misses' Goat Newport Ties,	"	864

ARTICLES.	BASIS.	QUANTITY.
Shoes, Misses' Kid Harvard Ties,	Pairs,	1,800
Shoes, Misses' Kid Newport Button,	"	2,160
Shoes, Misses' Kid Newport Ties,	"	9,000
Shoes, Misses' Kid Oxford Ties,	"	6,120
Shoes, Misses' Kid Button Sandals,	"	3,600
Shoes, Misses' Kid and Goat Ties,	"	3,600
Shoes, Misses' Kid Ties,	"	5,400
Shoes, Misses' Miscellaneous,	"	17,464
Shoes, Misses' Miscellaneous Hand-sewed,	"	13,500
Shoes, Misses' Patent Leather,	"	1,440
Shoes, Misses' Woolen and Felt,	"	3,600
Shoes, Misses', total,		72,940
Shoes, Women's Beaver Ties,		1,440
Shoes, Women's Glove Calf Ties,	"	2,160
Shoes, Women's Goat Ties,	"	4,680
Shoes, Women's Kid Ties,	"	23,640
Shoes, Women's French Kid Ties,	"	216
Shoes, Women's Sailor Kid Ties,	"	1,800
Shoes, Women's Miscellaneous Low Shoes,	"	157,200
Shoes, Women's Woolen and Felt Low Shoes,	"	3,600
Shoes, Women's Newport Button Glove Calf,	"	720
Shoes, Women's Newport Button Goat,	"	8,712
Shoes, Women's Newport But. Goat and Kid,	"	14,400
Shoes, Women's Newport Button Grain,	"	1,800
Shoes, Women's Newport Button Kid,	"	166,680
Shoes, Women's Newport Button India Kid,	"	18,000
Shoes, Women's Newport Miscellaneous,	"	324
Shoes, Women's Newport Ties, Glove Calf,	"	720
Shoes, Women's Newport Ties, Goat,	"	2,880
Shoes, Women's Newport Ties, Grain,	"	360
Shoes, Women's Newport Ties, Kid,	"	56,880
Shoes, Women's Newport Ties, India Kid,	"	18,000
Shoes, Women's Newport Ties, Miscellaneous,	"	10,440
Shoes, Women's Oxford Ties, Goat,	"	3,672
Shoes, Women's Oxford Ties, Glove Grain,	"	7,200

ARTICLES.	BASIS.
Shoes, Women's Oxford Ties, Kid,	Pairs,
Shoes, Women's Oxford Ties, French Kid,	"
Shoes, Women's Oxford Ties, Patent Leather,	"
Shoes, Women's Oxford Ties, Velvet Vamp,	"
Shoes, Women's Oxford Ties, Velvet,	
Shoes, Women's Ties, Glove Calf,	
Shees, Women's Ties, Wellesley,	
Shoes, Women's, total,	
Shoes, Youths' Miscellaneous,	
Shoes, Youths', total,	"
Slippers, Boys' Buck,	"
Slippers, Boys' Goat,	"
Slippers, Boys' Kid,	"
Slippers, Boys' Miscellaneous,	"
Slippers, Boys' Patent Leather,	"
Slippers, Boys' Turned,	"
Slippers, Boys' Velvet,	
Slippers, Boys' Velvet Pattern,	
Slippers, Boys', total,	"
Slippers, Children's Buck,	"
Slippers, Children's Glove Calf,	"
Slippers, Children's Hand-sewed,	"
Slippers, Children's Kid,	"
Slippers, Children's Miscellaneous,	"
Slippers, Children's, total,	
Slippers, Men's Alligator,	
Slippers, Men's Alligator Imitation,	"
Slippers, Men's Brocade,	"
Slippers, Men's Buck,	"
Slippers, Men's Calf and Goat,	"
Slippers, Men's Goat,	"
Slippers, Men's Grain,	"
Slippers, Men's Hair Calf,	"
Slippers, Men's Hand-sewed,	"

ARTICLES.	BASIS.	QUANTITY.
Slippers, Men's Kid,	Pairs,	3,600
Slippers, Men's Fancy Leather,	"	1,440
Slippers, Men's Opera Alligator Imitation,	"	5,400
Slippers, Men's Opera Goat,	"	13,320
Slippers, Men's Opera Leather,	"	13,716
Slippers, Men's Opera Low Cut,	"	103,410
Slippers, Men's Opera Patent Leather,	"	2,520
Slippers, Men's Pumps,	"	4,800
Slippers, Men's Turned,	"	12,000
Slippers, Men's Velvet,	"	93,096
Slippers, Men's Everett Velvet,	"	720
Slippers, Men's Opera Velvet,	"	8,496
Slippers, Men's Velvet Pattern,	"	2,160
Slippers, Men's, total,	"	501,982
Slippers, Misses' Alligator Imitation,	"	360
Slippers, Misses' Glove Calf,	"	120
Slippers, Misses' Goat,	"	8,280
Slippers, Misses' Grain,	"	180
Slippers, Misses' Kid,	"	12,780
Slippers, Misses' Kid Opera,	"	91,692
Slippers, Misses' Miscellaneous,	"	8,865
Slippers, Misses', total, ·	"	122,277
Slippers, Women's Beaver Croquet,	"	2,520
Slippers, Women's Croquet,	"	8,640
Slippers, Women's Fancy,	"	12,960
Slippers, Women's Glove Calf,	"	4,440
Slippers, Women's Glove Calf Opera,	"	1,080
Slippers, Women's Goat,	"	31,725
Slippers, Women's Goat Opera,	"	16,200
Slippers, Women's Goat Pointed,	"	1,800
Slippers, Women's Goat and Kid,	"	51,480
Slippers, Women's Goat and Kid Opera,	"	1,296
Slippers, Women's Glove Grain,	"	7,200
Slippers, Women's Kid,	"	317,400
Slippers, Women's French Kid Opera,	"	445,104

ARTICLES.	BASIS.	QUANTITY.
Slippers, Women's Kid Opera,	Pairs.	27,000
Slippers, Women's Patent Leather,	"	1,800
Slippers, Women's Patent Leather Opera,	"	10,800
Slippers, Women's Sandal,	"	64,000
Slippers, Women's Serge,	"	2,160
Slippers, Women's Turned,	"	36,000
Slippers, Women's Velvet,	"	22,320
Slippers, Women's Velvet Opera,	"	2,016
Slippers, Women's Miscel's Hand-sewed,	"	71,800
Slippers, Women's, total,	"	1,139,741
Slippers, Youths' Buck,	"	7,200
Slippers, Youths' Miscellaneous,	"	4,755
Slippers, Youths' Velvet Pattern,	"	360
Slippers, Youths', total,	"	12,315
Slippers, Boys', total,	"	34,212
Slippers, Children's, total,	"	105,324
Slippers, Men's, total,	"	501,982
Slippers, Misses', total,	"	122,277
Slippers, Women's, total,	"	1,139,741
Slippers, Youths', total,	"	12,315
Slippers, total,	"	1,915,851
Shoes, Boys', total,	"	4,752
Shoes, Children's, total,	"	44,520
Shoes, Men's, total,	"	787,880
Shoes, Misses', total,	"	72,940
Shoes, Women's, total,	"	929,012
Shoes, Youths', total,	"	4,750
Shoes, total,	"	1,843,854
Boots, Boys', total,	"	19,224
Boots, Children's, total,	"	112,056
Boots, Men's, total,	"	1,114,740
Boots, Misses', total,	"	410,160
Boots, Women's, total,	"	2,032,122

ARTICLES.	BASIS.	QUANTITY.
Boots, Youths', total,	Pairs,	17,672
Boots, total,	"	3,715,974

Whole amount of Boots, Shoes, and Slippers made in the City of Haverhill in 1888 is : —

	BASIS.	QUANTITY.
Boots,	Pairs,	3,715,974
Shoes,	"	1,843,854
Slippers,	"	1,915,851
Whole Amount,	"	7,475,679
Soles, Cut,	"	1,947,780
Soles, Inner,	"	650,304
Soles, Inner and Outer,	"	395,950
Soles, Men's,	"	224,640
Soles, Misses',	"	16,632
Soles, Women's,	"	313,200
Soles, Miscellaneous,	"	3,438,978
Soles, total,	"	6,997,484
Stiffenings, Children's,	"	24,000
Stiffenings, Men's,	"	36,720
Stiffenings, Misses',	"	30,720
Stiffenings, Women's,	"	60,720
Stiffenings, Leather Board,	"	163,620
Stiffenings, Miscellaneous,	"	1,061,460
Stiffenings, total,	"	1,317,240
Stitching, Worth,	Dollars,	250,997
Sundries, Worth,	"	13,200
	"	264,197
Taps, Children's,	Pairs,	10,800
Taps, Men's,	"	11,232
Taps, Misses',	"	22,032
Taps, Women's,	"	38,232
Taps, Miscellaneous,	"	757,060
Taps, total,	"	839,356
Toplifting, Worth,	Dollars,	12,000
Work on Boots and Shoes,	"	1,800
		13,800

Various Things.

Shoes, although made abundantly and well, are not the only things well made in Haverhill. Skilled workmen find employment in many other industries, of which the most important are the manufacture of hats, of woolens, of paper, and of morocco.

The manufacture of hats is quite an industry, the pay-roll for operatives in hat factories amounting to over $200,000 per year. The value of the wool and wool stock annually used is $175,000; fur, $100,000; hat bands, $60,000; silk cord, $6,000; sweat-leather, $15,000; strawboard and paper, $5,500; spool cotton, $3,500; cotton cloth, $3,000; soap, $3,000; shellac and gum, $5,000; dye-stuff and drugs, $10,000; satin, $9,000; oil, $1,000; other supplies, $10,000.

There are three large manufactories, — those of the Haverhill Hat Company, W. B. Thom and Company, and J. P. Gilman's Sons, making over 111,000 cases of fur and wool hats annually, and giving employment to 375 men and 125 women.

Of these, the oldest is the Haverhill Hat Company, located on Fleet Street, near the City Hall, which was incorporated in 1871 with a paid-up capital of $50,000, with Eben Mitchell as president and Charles Butters as treasurer. The business was first established about 1850 by P. Berkley How and Eben Mitchell, who carried on the works separately for some years and then formed a copartnership under the style of How and Mitchell, leasing the building now occupied by the Haverhill Hat Company. During the last twenty years the business has undergone many changes in methods of manufacture, and in the quality and variety of the goods made. Formerly, from 1,500 to 2,000 cases of hats were made up in anticipation of the semi-annual sales in January and July, while at the present time and for the last ten years the factory has been running exclusively on orders, sample cases only being made to sell from. While the earlier manufacturers were very successful, the goods they made would have but small sale to-day, some four or five colors and perhaps twenty or thirty styles being all that were then required, while now twenty or more colors and two hundred and fifty different styles are made up for every sale. The Haverhill Hat Company has a wide reputation for excellent colors, acknowledged by dealers to be excelled by those of no other manufacturer.

The business now owned by W. B. Thom and Company, originally established in Ayer's Village, was removed to this part of the city in 1874. Its growth

may be inferred from the fact, that, while the original
factory was equal to seventy dozen wool hats per day,
the present plant could make four hundred dozen per
day of fur and wool hats of all kinds. The works are
located on River Street, and include five buildings,
containing some thirty-two thousand feet of floor
space, besides engine houses, boiler houses, store
houses, etc. Their goods, distributed by their New
York house, find a ready market in all parts of the
world.

Four woolen mills are practically associated with
Haverhill,—one in the city itself, owned by Stevens
and Company, of North Andover, and three others, the
Groveland Mills in Groveland, managed by the trustees
of the estate of the late E. J. M. Hale. The male
employees in these four mills number 354 and the
females about 280, with a pay-roll amounting to
$260,000 per year. The goods manufactured by
Stevens and Company are women's dress goods of
various kinds, amounting to about 20,000 pieces. The
Groveland Mills manufacture flannels, making about
60,000 pieces annually. The wool used by these mills
amounts to 2,400,000 pounds, with supplies and other
material valued at $100,000 per annum.

The manufacture of morocco is carried on by two
firms,—Kimball and Son, and Lennox and Briggs,
who give employment to about 225 hands, with a
yearly pay-roll of about $75,000, finishing annually
about one million skins, valued at $700,000. The
leather is of a superior grade, they making a specialty
of "Dongola finished," which is an article of great

durability and sure to hold color. Kimball and Son occupy three three-story buildings on Fleet Street, and another on Pleasant Street, employing 130 hands, and making about 750,000 skins yearly, valued at $500,000. Lennox and Briggs occupy a two-story building in the rear of Washington Square, and part of two other buildings, employing 95 hands, making about 250,000 skins annually, valued at about $200,000. These firms supply both the Boston and the local markets, the demand being so great that their factories are kept running throughout the year at their greatest capacity, both plants having been enlarged during the past twelve months.

The Haverhill Paper Mill was organized in 1883 with a capital of $50,000, and has a large plant on the Bradford side of the river for the manufacture of newspaper. This mill gives employment to 50 hands, with a pay-roll amounting to about $160,000 per annum, and there are used about ten million pounds of material annually. This concern also has a mill at Berlin Falls, N. H.

The plant of the Haverhill Iron Works is situated on River Street. The company which operates it was organized in 1881 with a capital of $20,000, increased in 1883 to $40,000. The capacity of the works has been several times increased the past few years, and the present business is double what it has ever been before. The plant includes a large two-story building, a foundry, etc., and turns not only all ordinary iron work, architectural iron pieces, heating apparatus, etc., but also the most intricate machinery that is used in

the factories and shops, besides ornamental iron work
of any kind.

Among producers of goods intended purely for
home consumption is the Haverhill Gas Light Company, chartered by a special act of the Massachusetts
legislature, February 12, 1853. Its capital is $75,000
with a par value of $50 per share. The company's
principal works are on Winter Street along the Boston
and Maine Railroad. They are supplied with side-
tracks for the receipt of coal and other supplies and are
furnished with all improvements for abundant and
economical production. Most ample provision for
storage purposes has been recently made by the construction of a gasometer on Hilldale Avenue of a
capacity of 400,000 cubic feet. The total storage
capacity of the gasometers now in use is 580,000 cubic
feet. During the year ending June 30, 1887, these
works produced 38,096,000 cubic feet of gas of 19 can-
dle power; the present daily capacity is 320,000 cubic
feet. The company operates fifteen miles of street
mains and has in use over seventeen hundred meters.
Its financial condition is prosperous. Its plant is
valued at $75,000, and, adding to this, money invested,
cash on hand, and supplies, it had, in 1888, assets
amounting, in round numbers, to $92,000. The only
liabilities are the capital stock, and there was, therefore, in 1888 a balance of profit of about $17,000.
During the past ten years the price of gas has been
reduced from $3 per thousand feet to $1.50, the present
price. The management of the corporation has so
conducted its affairs as to fully satisfy its patrons and

the public, its extension of mains and increase of business keeping pace with the constant growth of the city.

The Haverhill Electric Company was organized as a corporation under the general laws of Massachusetts, on the sixth day of December, 1888. Its capital is $85,000; the par value of its shares, $100. The electric station is a large and commodious brick building situated on Essex Street along the line of the Boston and Maine Railroad, and within a few hundred feet of the manufacturing center of the city. It is furnished with four arc dynamos having a capacity of 165 lights, two incandescent dynamos, two engines of 250 horse power and three boilers of 350 horse power. The company at present furnishes 650 incandescent and 80 arc lights, 35 of the latter being used for street lighting. Thirty-five miles of wire are employed for the distribution of electricity throughout the city. Seven and a half miles are used exclusively for public street lamps.

Because of the great amount of light machinery required for the manufacture of shoes, Haverhill's chief industry, and owing to the disposition shown by some of our manufacturers to establish factories at some distance from the steam power plants in the present shoe district for the purpose of securing improved accommodations, the company is making special preparations for furnishing power for manufacturing purposes. There are in use at the present time eleven motors. The two daily and two weekly papers are printed by power furnished from this plant. A committee of the Board of Trade has been appointed to investigate and

report upon the feasibility of converting the power of
Mitchell's Falls upon the Merrimack River into electric force for manufacturing uses.

The Haverhill Electric Company has every reason
to hope for success. It is on a paying basis, free from
debt, and controlled by some of the most active and
public-spirited merchants and manufacturers. Its
president is the president of the Board of Trade.
The directors are now making arrangements for more
than duplicating the capacity of the works to satisfy
the public demand both for arc and incandescent lighting, and contracts have been made with the city for a
large increase of arc lighting and there is every prospect of this system coming into general use.

The water supply of the city is furnished by the
Haverhill Aqueduct Company. This company is a
corporation organized under the laws of Massachusetts
in the year 1802. Its capital is $300,000, divided into
1500 shares. The sources from which the water supply is drawn are Kenoza Lake, Lake Saltonstall,
Crystal Lake, and Round Pond. They are all within
the territorial limits of the city. Kenoza Lake, Lake
Saltonstall, and Round Pond lie on the highlands east
of the thickly settled portion of the city. Crystal Lake
lies west of the city proper. All these bodies of water
are deep and clear and are fed largely by springs.
Their waters are absolutely free from all obnoxious
vegetable matter and have been shown by frequent
analyses to be of remarkable purity. For many years
after the organization of this company its operations
were of necessity of the very simplest. Haverhill was

then a village of a few hundred inhabitants. Round Pond was then the only source of supply, the water being drawn from it through log pipes and distributed simply by force of gravitation. The increase of water facilities has, however, kept pace with the rapid growth of Haverhill, and the company has now under its control 40 miles of substantial pipe and two water-towers, one near Lake Kenoza, 40 feet in height, 30 feet in diameter, and of a capacity of 212,000 gallons, and another on Silver Hill, 60 feet in height, 40 feet in diameter and of a capacity of 575,000 gallons. The top of each of these towers is 256 feet above the Merrimack River and the business portion of the city. These towers are supplied by means of two Worthington pumps, each of the capacity of 2,000,000 gallons in twenty-four hours. The areas of the bodies of water from which this supply is drawn and their altitudes above the Merrimack River and the business portion of the city are as follows: —

	AREA.	ALTITUDE.
Kenoza Lake,	234 acres,	112 feet.
Crystal Lake,	175 "	148 "
Lake Saltonstall,	41 "	118 "
Round Pond,	38 "	148 "

After the great fire in 1882 a special connection was made with the "high service," i. e. the supply furnished through the water-towers for fire purposes. This special fire service consists of a twelve-inch main running through the business portion of the city and

PUMPING STATION.

supplying the fire hydrants belonging to the city,
besides a number of reservoirs. No other connection
with this pipe is permitted. It is capable of throwing
sixteen streams at once over our highest business
blocks without the aid of fire-engines. By means of an
electric indicator the height of the water in the water-
towers is recorded in the city fire-engine houses and at
the company's pumping station. The city fire alarm is
also connected with the company's pumping station,
where the Worthington pumps are always ready for
immediate use. It is the duty of the company's engineer during the progress of each fire to keep the
water in the water-towers at a height sufficient for the
greatest demands for fire purposes. The city owns
and uses 150 hydrants for fire service, the water for all
of which is furnished by the Aqueduct Company free
of any charge or expense to the citizens.

The water supply for the City of Haverhill, for
domestic, mechanical, and fire purposes, has thus far
been so abundant that never have any restrictions on
the liberal and even wasteful use of water been called
for. The present water supply is sufficient for a city
of one hundred thousand inhabitants, even without resort to additional dams or the use of the large natural
storage basins in the vicinity of the lakes. Under the
present system of supply Haverhill is practically supplied with two aqueducts. Each side of the city has
its lake above the river level and also a capacious
water-tower. Should an accident happen to the works
on one side of the city, an ample supply could be obtained from the other until such time as repairs could

be conveniently made. Owing to the abundance of water and the advantageous location of its sources, the three thousand water services are furnished at rates as low as in any New England city. The present perfection of our water system is due to the fact that the Haverhill Aqueduct Company has spared no expense to make its equipments and capacity fully adequate to the demands of the rapidly growing city in which it is situate.

With the natural advantages afforded by the vicinity of the lakes, aided by the institution of the high-pressure service above referred to, the fire department of Haverhill is one of the most efficient, and, being thoroughly equipped, organized, and trained, is ever ready for service when called upon. The department consists of one hundred and forty-nine officers and men, and includes one chief engineer, four assistant engineers, seven foremen, seven assistant foremen, three engineers of steamers, three stokers of steamers, fifteen hook and ladder men and thirty hose men, two drivers of steamers, three drivers of hose wagons, one driver of a chemical engine, one driver of a hook and ladder truck, and one man who acts as spare driver.

The city has spared no expense to make the department efficient. The apparatus consists of three steamers (all of the Amoskeag make), one chemical engine, one two-horse hose wagon, two one-horse hose wagons, one supply wagon, one hook and ladder truck, one one-horse hose carriage, three hand hose carriages, two hand engines, one engineer's wagon, and one chemical and protective wagon combined, with forty-three thousand feet of hose and thirteen horses.

There are seven engine houses, five in the city proper, one at Rocks Village, and one at Ayer's Village. A fire alarm telegraph is connected with the different engine houses. The system at present consists of twenty-seven boxes, nineteen miles of wire divided into four circuits, one bell striker, seven indicators, eight gongs, a five circuit repeater, and one hundred and three cells of batteries.

The efficiency of the department is also increased by the fact that the fire alarm is connected with the pumping station, where, immediately after an alarm is given, the pumps are set in motion by the engineer to replenish the water drawn from the reservoirs of the high-pressure service. With these facilities and with the present organization of the department it is evident that a fire is not likely to make great headway in the city. One may infer the efficiency of the department and of the men comprising it from the following record of fires taken from the chief engineer's annual report: Wingate School, insurance $10,000, loss $68; Numbers 1 to 17 Essex Street, insurance $7,800, loss $21; Hilldale Avenue, insurance $2,400, loss $20; Park Street, insurance $3,000, loss $85.

This naturally suggests the subject of insurance. Of course the facilities for obtaining insurance in Haverhill are much like those of other places. Nearly all the American and foreign companies are represented, and the rates of insurance are in accordance with risk and hazard. It must be confessed, that, for five years past, the insurance business has not been a remunerative one for the insurers. The great fire

February 17, 1882, cost the insurance companies two
and a half millions of dollars, and the losses by fire
during the years from 1882 to 1887 were also dispro-
portionately large, but, since the high-pressure service
was introduced, and since the appointment of the building inspector and the increase of the fire department,
the losses to the insurance companies in Haverhill are
not more than in any other city of the size, as can be
shown by the above mentioned report of the chief of
the fire department.

The New England Exchange placed a very high
rate of tariff on Haverhill property soon after the fire
of 1882, but reduced it fifty cents on mercantile risks
as soon as the high-pressure service was introduced.
And now, the Exchange is willing to reduce the tariff
on any individual risk, if the owner will make certain
improvements, such as supplying the buildings with
automatic sprinklers, automatic fire alarms, and shut-
ters, and will use gas instead of kerosene oil. Indeed,
it only depends upon the insured to have his property
rated as low as in any city in the United States if he
will follow the suggestions made for protection against
fire by the New England Exchange. In fact, many of
our recently erected buildings, and the older ones as
well, have been supplied with the improvements
alluded to, so that the expense of insurance on these
buildings is about one third of the cost in other similar
buildings where the improvements have not been
made.

Since the fire by which the City Hall was burnt up,
the city authorities, in conformity with the wishes of

the Exchange, have increased the apparatus of the fire
department by the addition of a new and improved
truck, and, in order to make the *personnel* of the department more efficient, have decided to elect the chief
engineer to serve during good behavior instead of
subjecting him to the risks of an annual election. The
engines are to be more widely scattered by the erection of new engine houses, which will enable the
department to reach the suburbs in reasonable time.

Among modern conveniences which it is the privilege of Haverhill to possess and utilize is its street
railway system, and the facilities it affords alike for
business and recreation rank high among the advantages the city possesses. It appears, by the nineteenth
(1888) annual report of the Railroad Commissioners,
taken in connection with the last census, that the
Haverhill and Groveland Street Railway Company had
a greater mileage of track to each thousand of the population in the communities served by it than any other
street railway system in the commonwealth. By its
cars the greater part of the citizens of Bradford, Groveland, and West Newbury are enabled to conveniently
reach the markets of Haverhill, to the mutual advantage of buyer and seller. Its influence is also most
important and beneficial in leading to the building up
of the suburban portions of Haverhill. It has been
true in the past, that the city was too compact, alike
for health and beauty. This came about from the
unwillingness of its people to dwell beyond easy walking distance of their work. Now they are availing
themselves of this cheap and easy method of reaching

the vacant spaces beyond, which are fast being dotted with houses, combining the main advantages of the city and country. Others who already own houses in the compact part of the city, and so cannot without loss wholly remove from it, are yet glad to avail themselves of the street car service in the warm season by boarding their families at some point in the rural portion of Haverhill or in some one of the towns adjoining, from which they can easily reach the center of business in the city and return at night, or earlier, to their families.

The officers of the company have, from the first, made special efforts to run cars at such times as to best accommodate the working people, thinking that the claims of those who are regular patrons and dependent upon their daily labor are first to be considered. At morning, noon, and night as many as ten cars, and often more, run to and from the shoe manufacturing district, almost or quite empty one way, and carrying operatives almost exclusively the other way.

In the season when those whose means and business permit it abandon the city for the pleasures and relaxations which summer resorts afford, the "stay-at-homes" find relief from heat and weariness on the open cars which bear them out in a few minutes to the heights overlooking the valley of Little River and the charming country beyond, or along the Merrimack valley amid scenery which has furnished themes and inspiration alike for poet and artist, and, better yet, has afforded year after year to thousands of the toil-worn such pleasure as neither poet nor artist could give.

From the Silver Hill terminus of the street railway,
Head's Hill in Bradford, with the river expanding into
the semblance of a lake at its base, are seen to good
advantage. For a considerable part of the distance
between Haverhill and the village of Groveland the
highway is so near the river that the passengers on the
open cars can watch with ease the various crafts which
at that season abound upon the noble river, and enjoy
the cool breeze which almost always tempers the heat
along its shores. From the substantial and nearly new
iron bridge over the Merrimack at Groveland a fine
view up and down the river is obtained. Beyond that
point, the highway in which the tracks are laid is at a
greater distance from the river, which, however, comes
into view for short stretches all through the ride to
West Newbury. A more charming picture than that
made by Rocks Village and the bridge with their
environments, as seen from the westerly part of the village of West Newbury, it would be hard to find in the
lower Merrimack valley.

In speaking of the street railway, one familiar with
its history must always call to mind, with deep regret
for his untimely decease, the late George W. Duncan,
without whose persistent efforts, it is safe to say, Haverhill would have had no street cars for at least five and
probably ten years later than the time (1877) when they
were introduced. At that time it was much more difficult to raise twenty-four thousand dollars in Haverhill
for any purpose than it would be to raise a hundred
thousand dollars now. And there were practically none
at that time who believed a street railway anywhere in

Haverhill would pay. That it did pay moderately from the first was due, in part, to exceptionally favorable circumstances. It was a line of only three miles in length, connecting the considerable village of Groveland with the business center of Haverhill over a practically level road. Still, it would have been easy, in spite of those advantages, to operate the road at a loss, and that result would probably have followed but for the careful management of its first directors, the Hon. Jackson B. Swett, the Hon. Levi Taylor, James D. White, Eben Mitchell, and George W. Duncan, the latter having also, as treasurer, the general management of the business.

From the small beginning in 1877, with only four cars and eight horses, it has increased until in 1888 it had thirty-eight cars, eighty-five horses, and a capital stock of $144,000, representing money actually paid in, principally owned in Haverhill, and being a larger amount than that invested in any other single business enterprise in the city, except, perhaps, that of supplying it with water, and possibly the flannel manufacturing business of M. T. Stevens and Company.

The commercial facilities of Haverhill are as good as can be desired and include direct transportation both by rail and by water from all points. The great Boston and Maine Railroad, which has arms extending in all directions, has three freight and three passenger depots within five minutes' walk of the heart of the city. From these more than twenty-five freight and seventy-six passsenger trains arrive and depart every twenty-four hours. In addition to this, the city has been made a

billing point within the past year, thus saving thousands
of dollars to shippers annually. Haverhill is also at
the head of navigation of the Merrimack River. From
this point to the sea, a distance of sixteen miles, the
channel of the river is broad and deep. More than a
hundred schooners and a large number of coal, granite,
and lumber scows arrive at this port every year, and
their cargoes are delivered directly to the business
localities. The river is of indirect though none the less
real value in serving, by the opportunities for competition it affords, to keep railway rates for freight at a
reasonable figure.

Brick making began in Haverhill more than two
hundred years ago, when the husband of the heroic
Hannah Duston was guarded by a file of soldiers as he
brought the clay from the pits to the yard near his
house. Ever since that eventful period Haverhill has
not only supplied its own bricks, but large quantities
are also sent to Lawrence, Lowell, and other cities and
towns. The clay pits are situated about a mile and a
half north of the city near the railroad, and the material
is the best in color and strength to be found in New
England. With the opportunity of taking the bricks
from the yard directly to the building sites in a half
hour, and in unlimited quantity, it is safe to assume
that Haverhill will always be able to secure this essential element of substantial growth at as low price as
any city in the country. Within the city limits is also
a fine granite yard, while the opportunities for bringing
granite to its very doors by the Merrimack River are
unsurpassed. Lumber and all kinds of wood building

material are also abundant, and three extensive and
growing firms supply everything that is needed in that
line.

As can readily be seen from these facts, the strongest inducements are offered to prospective builders of
business blocks, while a house and lot complete, suitable
for any man with a small income, can be put up for
from one thousand to twelve hundred dollars, and this
on the line of the horse railway and within a ride of
from five to ten minutes of the business center of the
city.

The newspapers of Haverhill consist of two daily
and two weekly issues, which find a large circulation
in the adjacent Massachusetts and New Hampshire
towns as well as in the city itself. There have been,
from time to time, other ventures in the field of journal-
ism, but the final result for the present seems to suggest
the survival of the fittest.

The Daily Bulletin was started July 1, 1871, and
is therefore the oldest daily paper in the city. Its
publication was begun in the face of great obstacles
and with many predictions as to its ultimate and speedy
collapse. In fact, with so little favor was the scheme
of a daily paper in this city viewed that only about one
hundred and fifty subscribers could be obtained. For
five years the paper struggled for existence, but September 17, 1875, the present proprietors, I. L. Mitchell
and Warren Hoyt, bought out the original proprietor,
Mr. A. J. Hoyt, and in 1877 the Tri-Weekly Publisher
was bought and merged with it. Since that time the
growth of the paper has been gradual and steady. Year

by year it has strengthened its hold upon the public, until, today, it stands among the leading dailies in Essex County. For the first seventeen years the office and plant were at No. 4 Main Street, although its increasing growth compelled the enlargement of the establishment before the paper was a decade old. In 1888 new quarters were obliged to be sought, owing to the fact that additional room was required for both editorial and job departments. On April 5 of that year the establishment was removed to the Daggett Building, in which structure the Bulletin now occupies three floors. It boasts at the present time one of the most centrally located, most convenient, and thoroughly equipped establishments in this section of the state. The politics of the paper have always been Republican, but the aims of its proprietors have been toward independence rather than ultra-partisan. The paper is also essentially a local sheet. Its aim is to cover Haverhill and vicinity thoroughly, and, while attention is given to general news, yet local news is considered of the first and greatest importances. In connection with the paper is a large book and job printing establishment in which skilled help is employed the year round and which has facilities for all kinds of fine work.

The Gazette goes back to very early times in the history of Haverhill, it having been established in 1798, though the daily edition was of much later origin. It publishes now both weekly and daily editions, the latter printed on a double cylinder Hoe press. The Gazette has a wide circulation, and is a bright, interesting, and influential paper. Connected with the establishment is

a large job and book printing office, where is printed
the Popular Science News and Boston Journal of
Chemistry.

The people of Haverhill are an amusement loving
and an amusement enjoying class. The supply is almost
always equal to the demand, especially in a case of this
sort, and in consequence Haverhill is well provided
with places where its hard-working citizens can obtain
rest and enjoyment when the labors of the day are over.
First and foremost among these is the Academy of
Music, one of the prettiest, best arranged, best equipped,
and largest theaters, outside of Boston, in New England.
Here are presented the best dramatic attractions on the
road; and during the season, which extends from September to June, all the stars in the dramatic firmament
shine before the people. Manager James F. West
exercises good judgment in securing talent, and, although
the range of attractions is large, including comedy,
tragedy, variety, opera, both light and heavy, concerts,
both vocal and instrumental, and those nondescript
plays, neither one thing nor the other, but which might
be included under the head of farces, yet only the best
under that head are booked. The average is about two
performances a week, and hence, as may be readily
seen, as far as theatrical performances are concerned
no place in America of its size is better supplied. The
names of Booth, Barrett, (Lawrence and Wilson,) Keene,
Dowling, Mrs. Langtry, Julia Marlowe, Georgia Cayvan,
Joseph Jefferson, Margaret Mather, Fanny Davenport,
Modjeska, Janauschek, Rhea, Lotta, Annie Pixley,
Denman Thompson, Gilmore's Band and Boston Sym-

phony Orchestra are not only familiar to Haverhill but they have been seen again and again upon its stage.

In addition there is never a season in which one or more courses of lectures and semi-private entertainments are not given. The city boasts talent and genius, fine musicians, good vocalists, amateur actors, and elocutionists, and they are never loth to respond to the calls made upon them for charitable and social purposes. Moreover the bazaar, fair, sale, epidemic under various disguises, rages as virulently in Haverhill as is possible, and their number is legion. Such are some of the amusements which attract and entertain our citizens in winter, to say nothing of skating rinks, sleighing parties, ice skating, either on the river or on the beautiful lakes with which the vicinity abounds, dances private and public, etc., but it is in summer that Haverhill affords amusements which far surpass those offered by the ordinary small city.

On the beautiful Merrimack River, which equals in clear, tranquil, calm beauty any similar river in this country, pleasure steamers ply, loaded with human freight, every pleasant summer day, bound either for the salt and invigorating breezes to be found at "Black Rocks," the Coney Island of New England, or else to find rest and shady coolness in the nooks and woody ravines of Eagle Island, The Pines, and Balch's Grove, public places for picnic devotees which lie along the ighteen mile stretch from Haverhill to the mouth of the river. Within a radius of twenty, nay ten, miles from the very heart of business life, over twenty lakes lie nestled among the green fields, surrounded by groves of large and beautiful trees. To these also during the summer months the seekers after rest and amusement make their way to picnic and enjoy the out-door sports of which Americans, especially Young America, are so fond. In summer also amateur base ball teams flourish, and on the large and well equipped grounds, known as "Recreation Park," furnish entertainment to many. The list might be continued indefinitely, for Haverhill boasts several lawn tennis clubs, two yacht clubs, a large number of amateur boatmen, hunters and fishermen galore, a good half-mile track on which meetings which draw out good exhibitions of speed are held, a rifle club, a bicycle club, an amateur photographers' club, etc., etc. In truth the opportunities afforded for amusement, no matter what the season of the year may be, are many and are enjoyed to their full extent.

The Kenoza Club, an association of gentlemen already referred to, has recently developed an access of energy and has added to its house on the edge of the lake from which it derives its name a large veranda and pavilion which handsomely equips it for social pleasures.

While it is unnecessary in Haverhill for a newcomer to be a member of some secret organization in order to receive cordial recognition and welcome, it should be stated that those belonging to almost any secret or social organization in existence will find societies ready to give them the fraternal sign and greeting. The first Free Mason's lodge was chartered in 1802. There are at present two lodges, a chapter of Royal

Arch Masons, a council of Select and Royal Masters, a Commandery of Knights Templar of 188 members, the Lodge of Perfection, Princes of Jerusalem Council, Rose Croix Chapter, and a Kadash Council. There are

ODD FELLOWS' BUILDING.

seven lodges of Odd Fellows with a very large membership, beside large orders of Knights of Pythias, Knights of Honor, Red Men, Pilgrim Fathers, Ancient Order of Hibernians, United American Mechanics, and many others, in all comprising thirty or more different organizations, some of them having fine club rooms as well as halls for business. In addition to these there

are two fine private social organizations, the Pentucket
Club and the Wachusett Club, each having most pleasant
and tasteful quarters which do much to add to the social
attractions of city.

A rich and extensive farming country depends
largely upon Haverhill for a market for its products of
the soil. Fresh supplies for the table can always be
found in abundance and at low prices. Rents vary
from eight dollars per month for tenements of five or
six rooms to fifteen dollars for those of the latest
modern conveniences, and whole houses rent for from
the latter figures to thirty dollars per month. Board
for mechanics costs from three to five dollars per week,
and at the hotels from six to nine per week. These
figures can only be given approximately, but, taking
into consideration the attractions and advantages of the
city, both natural and acquired, as a place of residence,
the cost of living is remarkably low. Mechanics in
many cases own their own houses and in all cases
they can do so. Haverhill has as many cozy little
homes owned by workingmen as any other city of its
size in the Union. This is largely due, not alone to the
encouragement given them to build by the public
spirited capitalist, but more especially to the two local
co-operative banks, which in other parts of the country
are known as building, loan, or savings associations.
Institutions of this kind are doubtless among the greatest boons of a private nature to working people that have
been offered them in this country. The two banks referred to are both in an exceedingly flourishing condition, having a large accumulated capital gathered
from the savings of working men and women.

A Place to Live In.

The beautiful situation of Haverhill upon the banks
of the noble Merrimack, the commanding heights upon
which our houses can be so built that almost all may
have magnificent views of the river valley and the
surrounding country, and also a flood of sunlight
and an abundance of pure air, form natural advantages
which few cities can boast, but which are by no means
all that we enjoy.

Far enough from the sea to have the raw east
winds somewhat tempered, near enough to the mountains to get their unadulterated health-giving air, there
is no blessing which the climate of New England can
give that is not ours. The elevation of the river banks
raises them from whatever danger might arise from
dampness, and affords admirable facilities for the best
drainage through a soil that has sufficient fertility and
is of such variety as to afford flourishing life to all the

beautiful trees, flowers, and vegetables, either native or imported, which thrive anywhere in New England. The fine shade-trees in almost all the streets occupied by residences form a marked feature of the attractiveness of the city, and one which is seldom found in a manufacturing community.

The four beautiful lakes, to the banks of which some of our most attractive building lots have been extended, offer, in addition to an abundant supply of pure water for all purposes, suburban walks and drives of unexcelled beauty.

In fact, the hills of Haverhill, especially those overlooking these lovely lakes and the glorious river, are among the most favored spots on earth for human residence, affording opportunities for the most delightful surroundings. Every acre is so situated that a desirable home may be made upon it, adapted to every taste in regard to altitude, grade, and exposure. The infinite variety of slopes to every point of the compass enables one to choose where the morning and the evening sun shall shine upon his house, whether he shall be protected from the north, the south, the east, or the west winds, or whether he shall welcome the breezes from every point.

No similar advantages does any other city in the country furnish within so short a distance from a common center. The incalculable blessing of such homes to the character of an entire community cannot be overestimated. The child brought up among such glorious surroundings cannot fail to be affected by their elevating influences, and must imbibe insensibly high, strong, and wholesome habits of thought. To the hard-worked man nothing affords greater relief, gives greater strength for the daily struggle, than the ability in one moment to turn his back upon the din and turmoil and dust and confusion, the inevitable concomitants of busy quarters, and from his hill-side cottage breathe the pure air of heaven, with one of the most perfect of earth's pictures stretched before his eye.

This is no imaginary sketch. Every man that can buy a house lot or that can pay rent has it in his power to choose one of these situations, instead of huddling close to his factory on the river bank, because he is too lazy or too indifferent to choose more wisely. Every inch of land in the lower levels of the city is none too much for its business uses, and, by the aid of the horse railway for the more distant parts, a large extent of our territory is made available for dwellings. Every house can be within easy reach of one or more of our fine bodies of water, affording delight to the eye as well as boating, fishing, and bathing facilities. The noble Merrimack, flowing at our feet, is no small item in the grand sum of benefits which nature has bestowed upon this spot, enabling us to reach the great ocean and by it all the ports of the world. The water in this river is deep enough to float to our wharves vessels larger than those employed in our merchant marine when this county led the whole continent in its foreign commerce. That the river can be utilized as a water power is the opinion of competent engineers, another gift of nature not to be overlooked.

The distance from Boston (to which it is near enough for the convenient transaction of the business which naturally gravitates to great centers, and from which it is far enough not to be absorbed as a suburb) is an advantage the importance of which can hardly be overestimated, enabling us to form a society sufficiently independent to have a character of its own, yet within such easy reach of cosmopolitan influences as to avoid all danger of provincialism. On our frequent trips to the metropolis, the beautiful glimpses of wood, meadow, lake, and river in the short hour's journey afford a pleasing variety which is an alleviation to the toil of the day.

But it is not to natural advantages alone that one looks when about to take up a new residence. Religion, the recognition of God as an object of worship, love, and obedience, the corner-stone on which our civilization rests, calling out as it does the best there is in us, must occupy a prominent place in every man's thoughts. Whatever form of Christian belief one may hold, he can be reasonably sure of finding some of his household of faith established in this city, ready to welcome him with kindly sympathy. In few communities does the religious spirit hold stronger sway, every year showing an advance in this direction, owing perhaps in large measure to the fact that in all sects religious worship has been freed from much of its old time austerity.

The opportunities for education are ample in almost every New England city, but here in Haverhill we are especially favored in our admirable educational advantages for both sexes and for all ages. We have not only our excellent public schools, at the head of which

stands a high school at which our young men are
fitted for college or for the duties of citizenship, but innumerable clubs and associations, having for their ultimate object the better education of men and women.
Our public school system, receiving the active and intelligent support of our best citizens on its committees, and
being peculiarly favored in its well-established teachers,
meets the approbation of all, and the results achieved
by it are eminently satisfactory. Our private schools,
beginning with those for children of the tenderest years,
are conducted on the best plans, instilling ideas and
principles which it was once thought could be obtained
nowhere but at home. In this connection we must not
forget the close proximity of the Bradford Academy for
girls, which has almost a national reputation, and an
excellent private school for boys, just across the river.

The old-fashioned lyceum seems to have ceased to
exist, but in its place we have numerous literary clubs
which are often instructed by the best talent in this
country or perhaps in the world; and under the auspices
of our various societies, notably the Young Men's Christian Association, lectures and other instructive entertainments afford ample opportunities for mental improvement. Greater facilities are now being offered
for our musical education, which has hitherto been
somewhat neglected, and we hope soon to furnish ap-
preciative audiences for the encouragement of the best
music, which is always at our command. The drama
in a sufficiently elevating form to have an educational
influence can hardly be said to have gained a permanent foothold with us, notwithstanding the ample facili-

ties furnished by the able management of our beautiful
Academy of Music, but we hope, as we progresss in
wisdom and prosperity, soon to add this to our privileges.

Drawing, painting, and even sculpture have their
part in our schools, and, together with classes especially
devoted to these branches of fine art, have succeeded
in developing talent of which we have reason to be
proud. That we show a keen appreciation of good work
is the verdict of some of the first artists in the country. As a powerful instrument for intellectual improvement and recreation, we have a public library, well
endowed and admirably conducted. According to the
report of the commissioner of education there are but
nine free lending libraries larger than ours, which contains forty-five thousand volumes. Physical education
is receiving more attention, as the establishment of an
excellent gymnasium with competent teachers, in connection with the Young Men's Christian Association,
attests; and the numerous ponies with children on their
backs in our streets show that the important branch of
horsemanship is not neglected. Dancing schools have
also been established and are well patronized. In considering the social life of Haverhill, it can be said by
the writer, that there is no place it has been his good
fortune to visit, in a somewhat extended experience of
towns in this country, where so cordial a welcome is
extended to the new comer, where a man so instantly
finds himself in possession of all the privileges which
are often obtained only at the expense of long residence. He can speak from his own experience and

that of every adopted citizen, who will join heartily in
this expression. It is impossible to say too much of
the hearty good-will and kindliness of spirit which greet
every man, woman, and child who enters the arena in
whatever capacity, provide a fair field for the exercise
of every talent, and aid every laudable effort however
humble.

That we are hedged in by no artificial barriers is
one of our greatest blessings, and one which more than
anything else perhaps invites accession to our number.
If we do not as fully as we ought avail ourselves of the
privileges of mutual improvement and social enjoyment, it must be laid to our too great devotion to business. For some years we were able to point out to
the stranger our one gentlemen of leisure, but he has
long since joined the great army of workers, finding, presumably, his isolated position insupportable.
There is no reason why Haverhill should not afford,
however, a delightful residence for gentlemen of leisure,
but business strife is so hot they seem to have found no
place so far. One may reasonably look forward, however, to a sufficient cessation of this busy life as to enable us to test the admirable material we have for social
enjoyment. One pleasing feature is gaining daily
prominence and will prove a great benefit to us, viz.,
the increase in the number of social meetings of employers and employed. Nothing can add to the general
solidity of a town so much as these pleasant and cordial relations.

In an article recently published in one of the newspapers of the city, after mentioning the beauty, the

health (indicated by the bright, animated looks, quick, independent walk, and general air of happiness), and the taste in dress of the women seen on our streets, the writer goes on to say of the men: "There is a brightness, an animation, an expression of shrewdness visible upon the lineaments of every passer-by which speaks volumes for its possessor's brain, mind, and soul. Moreover, these characteristics are inherent in most of the operatives in this city. They are superior in intellect, general knowledge, and schooling to any similar class in America. They are thinking, reasoning men, strong in their convictions, outspoken in their opinions, strong in the faith inherited from sturdy, independent ancestors." Formed of such elements, the social fabric of Haverhill should be strong. The man who was yesterday employed is to-day an employer, as every avenue is open to energetic and intelligent action. Under a republican form of government, this may be said to be true of every city and town in the land, but every one knows that in many places local influences often handicap the ambitious aspirant. That the local influences here all favor the man who tries to rise is what the writer desires especially to emphasize.

The natural and acquired advantages of Haverhill have already been frequently alluded to, and it remains here but to touch upon the use that may be made of them in relation to business. That the situation of our beautiful city is thoroughly advantageous for the transaction of almost any kind of business has been pointed out. The fine sites for factories, extending for nine miles on the banks of the Merrimack and to the New

Hampshire line in the Little River valley, with all the advantages of river and railroad transportation, the healthful surroundings without which successful work is impossible, the formation of the land, enabling us to live within easy rich of our factories and yet in a different atmosphere, all go to make up a sum total of inestimable value. Our religious, educational, and social privileges all have immense weight in the business world, and, by their influence on our citizens, become active agents in the promotion of every enterprise. Every business man knows the value of intelligent, educated, skilled workmen, and what a vast difference there is in the conduct and success of an establishment where these can be obtained, and one where ignorant labor is employed. Nowhere is this phase more propitious than here.

A larger question, and one of greater import in the long run than the mere question of labor to the man planting his business here, is that the whole conduct of the affairs of the city by the selection of its officers is in the hands of an intelligent people who make Haverhill their permanent home and do not leave us at the mercy of a shifting population. The latter is often the case in manufacturing towns where foreign capital alone is invested. We are fortunate in that our citizens make and spend their money here. The stranger is at once impressed by our elegant and comfortable residences, so superior in number and beauty to those of other cities much larger, where prosperity is less generally diffused. This is our strong point, that we are a homogeneous household, depending upon each other and

absolutely controlling our own affairs. If this is not a
community which invites accessions, where can one be
found?

Our building facilities are unexcelled. The best of
building stones, especially for foundations, can be bought
for little more than the expense of hauling, as our hill-
sides are full of them. Good bricks are made from the
best of clay within our borders so cheaply that all the
neighboring cities are supplied by us. The river enables us to bring timber and lime to our wharves at
reasonable rates. Our iron works furnish everything
of machinery and heating apparatus, in successful competition with the largest establishments in the country.
Our hardware stores supply all the materials in their
line at wholesale prices. So that buildings can be
erected and equipped here to the best advantage.

An instance of the latest building enterprise is seen
in the handsome Daggett Building, which towers above
Merrimack Street and rivals in its appointments metropolitan edifices.

For the prosecution of business the same advantages
apply. At no inland town can coal be furnished so
cheaply. River transportation and wharf privileges
enable us to procure all the more bulky articles, such
as the timber, iron and other metals that go into many
branches of manufacture, moulding sand, granite, oil,
tar, even cotton and wool, at rates which our railroads
are compelled to meet.

Rents are reasonable, and the co-operative banks
furnish the means for the easy acquisition of homes;
our real estate owners favor the establishment of

DAGGETT BUILDING.

homes; no land is held at fancy prices; the position
of landlord is not sought: no place that is worth living
in offers greater inducements for householders. The
cost of living is not excessive, although the general
prosperity has created a demand for the best the market
affords, and consequently has somewhat enhanced prices
over those of more stagnant communities.

No manufacturer ever left Haverhill or ever will
leave it except for the one expected advantage, cheaper
labor, as the questions of rent, power, and taxes are
entirely subordinate and are manifestly counter-balanced
by others; and, as cheaper labor has been found to
result in a product of lower standard, it is only a question of time and the action of the natural laws of
demand and supply, untrammeled by artificial condi-
tions caused by unhealthy agitation, when our city, as a
center of skilled labor, will inevitably recall the wan-
dering ones whose hearts are still with us, together
with an army of new recruits. Haverhill stands ready
to welcome all.

Our banks especially favor the business men of
Haverhill and seek no outside loans until every citizen
who by his character shows himself worthy of credit is
accommodated. The character of our workmen has
been mentioned, but that we have within easy reach
five thousand men in addition to our own population of
twenty-six thousand is a fact worth considering, especially in view of the enlargement of the shoe business,
as most of these men are skilled in that craft. But it is
not to the extension of the shoe business alone that we
look; believing that a diversity of industries is advantageous to a community, we offer inducements to all.

Our retail stores, supplying a large surrounding country, are admirably conducted by enterprising men, and no one need seek elsewhere for the gratification of any reasonable want in their line. Our advantages might be enlarged upon almost indefinitely, but the scope of this paper is merely to mention some of the most marked, confidently trusting that they will arrest the attention of outsiders. Let us not forget, that, while furnishing opportunities for the strong in mind and body, the community is not unmindful of those who are disabled by accident or ill health, who are cared for in our well appointed and ably managed City Hospital, and that the poor and unfortunate are so wisely assisted by our benevolent institutions, the Old Ladies' Home, the Children's Home, the Benevolent Society, etc., as not to create paupers, who are consequently few in number.

PROMINENT

Business Interests

of

HAVERHILL.

N. F. Sawyer.

Mr. N. F. Sawyer, whose shop is in the rear of 72 Washington Street, is the manufacturer and patentee of the most powerful heater for both steam and hot

water heating yet invented, which possesses more heating surface which comes in direct action with the fire than any other, and for hot water heating has the best water circulation of all, being free, rapid, and positive.

Edgar O. Bullock.

Who for 18 years had been connected with dry goods houses in Boston, formed in 1882 a partnership with O. W. Butters, then doing a business of $20,000 a year in the cutting of shoe stock. In 1885 Mr. Butters retired and Mr. Bullock has continued alone. He occupies the whole building at 45 and 47 Washington Street and the upper floor of the next building. He

does a business of $120,000 a year, employs twenty-five hands the year round, cutting over a ton of leather a day. Most of this comes direct from the tanneries, making a saving in freight and securing a uniform line of stock. This, with new and improved machinery, careful handling, and a perfect system in the factory, produces goods that command a ready sale and good prices.

C. N. Rhodes.

Mr. C. N. Rhodes, a large dealer in ladies' furnishing goods, millinery, domestics, oil and straw carpetings, rugs, and Buttrick's patterns, at Nos. 52 and 54 Merrimack Street, began business in 1865 at No. 10 Main Street, occupying for two years one floor, the two years following two floors, while at the end of the fourth year the demands of his business for space required the whole building. After about eight years he removed to the Odd Fellows' Building, No. 28 Main Street, remaining there about nine years, whence he removed to his present large and commodious store, which has a floor surface of over forty-two hundred square feet. In accordance with the requirements of a large business at the present day, he early adopted the cash carrying system, using for four years the Lamson ball system, and for the next four years the Lamson wire system.

Starting at the close of the great War of the Rebellion, when everything had a fictitious value, the prices of merchandise have decreased almost continually up to the present hour. For example, imported corsets, which were then sold at retail at three dollars and a half a pair, now pay a profit at ninety-two cents. Spool cotton sold then for fifteen cents, and sells now for two cents. Forty-inch sheeting, which then sold for seventy-five cents, sells now for eight cents. Yet Mr. Rhodes' increase in trade, as measured by the receipts, has more than kept pace with the fall in prices; and now the services of from ten to eighteen clerks are required.

S. A. Dow.

Mr. S. A. Dow began business in a small way in this city in the year 1883, engaging in the sale of pianos, organs, musical instruments in general, rich stationery, bric-a-brac, and so forth, but now, in contrast to this small beginning, is doing the largest business in this line in the city of Haverhill to-day. He occupies the store No. 85 Merrimack Street, which has been fitted

up purposely for his occupancy. On the first floor is his salesroom, which is very handsomely equipped. On the second floor he has a large studio, while in the rear is the framing department, in which only work of the best quality is done. Mr. Dow is the agent for some of the best musical instruments in the world, notably the Henry Miller, Behr Bros., Newby, and Evans pianos, the Mason and Hamlin, Estey, and Sterling organs.

J. C. Hardy.

Mr. J. C. Hardy is the proprietor of a flourishing and constantly increasing business in grain, hay, straw, flour, coal, and wood. His warehouse, forty-five feet by seventy-five, is a brick building, built by himself in 1870. It is located at No. 188 Winter Street, on the line of the Boston and Maine Railway, in a situation convenient alike for dealer and customer. It has a

cemented cellar and possesses the very necessary quality of dryness, so much so as to fit it for a storehouse for grain. Mr. Hardy ships his hay and straw from New Hampshire, Maine, Canada, and New York, while his flour and grain he brings in directly from the West. He received last year about one thousand car-loads of merchandise, and handled five hundred tons of hay and about three thousand tons of coal. Since April, 1887, he has occupied No. 8 Emerson Street as a branch store.

George H. Hill.

Twenty years ago the subject of this sketch began on a small scale, in connection with his father, C. H. Hill, who kept a grocery store at 108 Winter Street, the business to which, in later years, he devoted his entire energy and time. His original stock consisted of a few potted plants which were sold in connection with the store goods. As the demand increased the stock in trade enlarged until in a few years the business had grown to such proportions that he leased a store, 44 Winter Street, and devoted his entire time to the sale of plants and flowers. The limited accommodations here soon necessitated another change, and in 1885 the store at 14 Winter Street was fitted up and filled with a select and ever increasing stock of flowers, flowering plants, and ornamental shrubs. Here are to be found at all times the rare novelties and newest varieties of the floral creation, and work from this establishment is justly celebrated. Mr. Hill is, and has been, closely identified with the rise and growth of floral culture in Haverhill. Twenty years ago not one well laid out or one well kept lawn could have been found within our city limits. Scarcely a house could be found that could boast of a well kept flower garden, while ornamental trees and shrubs were practically unknown. Now all this is changed, and Haverhill homes are noted for their beautiful surroundings. To Mr. Hill and his efforts is due in a great measure this marvelous change in public taste and opinion, and from his long experience he is able to give ideas in floriculture that must be of value to his patrons and the public.

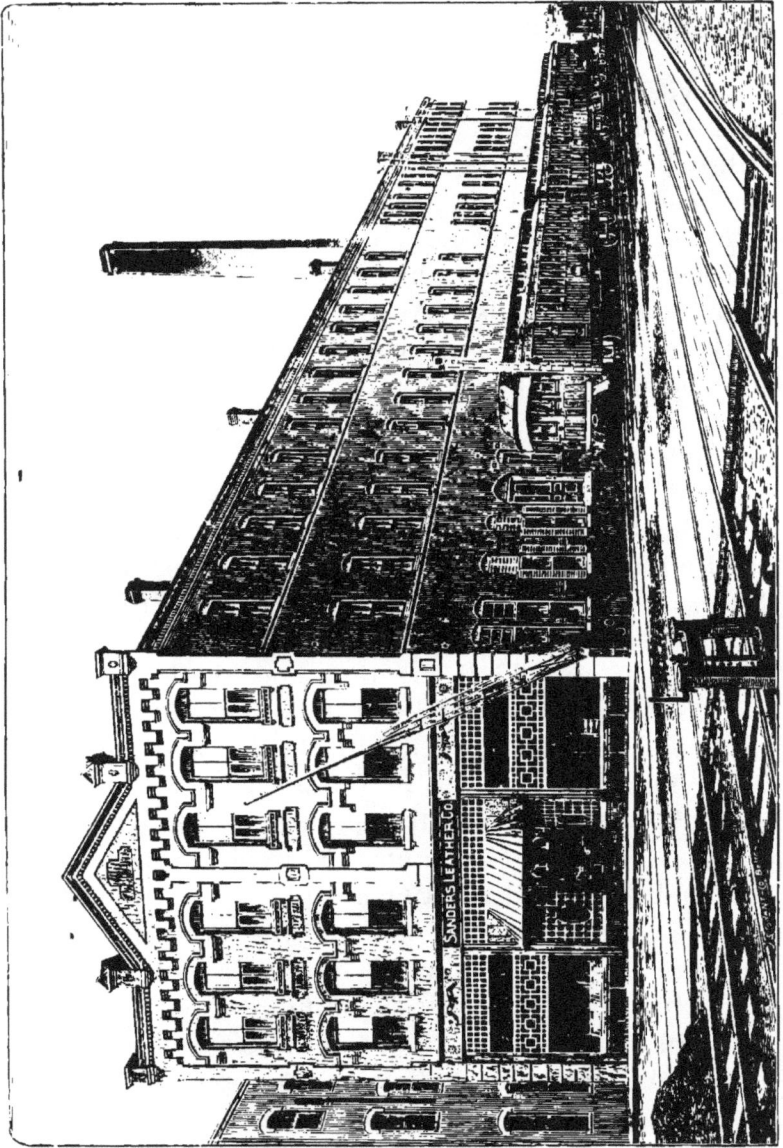

The Sanders Leather Company.

Prior to 1870 every boot and shoe manufacturer was obliged to buy his sole leather by the side and to devote a large part of the room and labor of his factory to cutting and sorting it. This was a great disadvantage to him, as not only was a considerable amount of capital and labor involved, but, owing to the innumerable grades and qualities in a side of leather, he found himself loaded with a large proportion which he could not use.

Recognizing that in the numerous special lines of manufacture in this city there was a demand for every part of the leather if each could be put where it belonged, Mr. Thomas Sanders in 1870 started the business of sole leather cutting on a large scale, driving the entering wedge which has since revolutionized the system of manufacture in this city.

The Sanders Leather Company which succeeded to this business in 1883, is still managed by Mr. Sanders, its president, and has steadily done a business of half a million dollars a year. In 1889 a considerable addition has been made to the facilities of the company, which will enable it to do a business of three quarters of a million in future, cutting about 4,000 sides a week of the best union and oak leather. The business has extended to all parts of the United States where boots and shoes are made, very few enlightened manufacturers adhering to the old system of cutting their own leather.

Many of the largest manufacturers in the West and South are the regular customers of the Sanders Leather Company.

Chase Brothers.

This firm of manufacturing stationers is composed of Messrs. George F., and Herbert A. Chase, both young men, who started a small printing business January 6, 1878, with one press, doing all the work themselves, since which time they have steadily enlarged to meet the demands of their increasing trade, until in 1889, the plant in their printing department includes six presses of the most approved patterns, together with all the standard faces of type and every necessary appliance for the rapid production of first class work of every description. In connection with the above is a blank book manufactory, and a stationery department where can be found every variety of blank

ONLY PRESS 1878.

ONE OF SIX PRESSES 1889.

The Haverhill Bindery.

books, office and counting room supplies: a feature
of the business being the manufacture to order of
special blank books, this being the only manufactory in the city. From their small beginning eleven
years ago, the firm now occupies the four story brick
building, Nos. 13 and 15 Washington Street. The first
book ever published, printed, and bound in this city
came from this establishment.

Previous to January, 1887, there was no book
bindery in this city, and it was necessary to send all
work out of town for binding.
Messrs. Chase Brothers, realizing that this caused many
delays and was a great inconvenience to their customers,
added this department to their
business, with the intention of
doing only their own binding,
thus having all work under
their immediate control and
supervision. That this enterprise was appreciated is
shown by the fact that their
order trade has more than
doubled since the addition, and a large and increasing
business comes from out of town. In this department
are manufactured the Excelsior blank books, which
are recognized as the most complete line in the trade,
the ledger paper being manufactured especially for
them, and each book receives a custom binding far
superior to the "team work" on many competing lines.

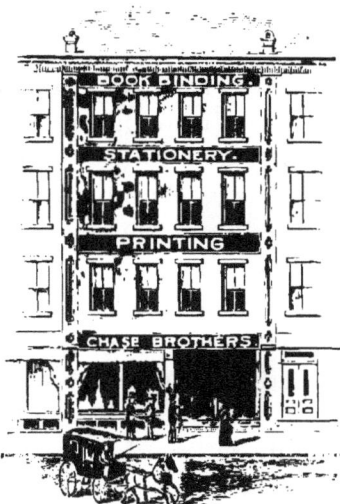

The Carleton School.

The village of Bradford, opposite and within easy reach of Haverhill, has always been a favored locality with regard to schools, from Father Greenleaf's time, when the celebrated Bradford Academy, then a school for both sexes, was under his guardianship, until the present, when side by side with this time-honored institution, now reserved for the gentler sex alone, stands another school, adapted for masculine youthfulness and vigor.

In the center of this healthful and beautiful village, and occupying its most attractive site, is the Carleton School. This institution was established in 1884 and is a classical and English school for boys.

The principal, I. N. Carleton, A. M., Ph. D. is well known as a former instructor for four years in Phillips Academy, Andover, and for fourteen years principal of the State Normal School of Connecticut, at New Britain. He is assisted by a well qualified corps of teachers, and is thus able to give to pupils the individual attention that they need, and which can not be obtained in a large school.

Parents traveling abroad, or for any other reason unable to provide a suitable home for their boys, can here find the comforts and advantages of a cultivated home and a thorough school, besides those naturally attached to a quiet village which is yet within a moment's reach of a large city. The disposition of the individual scholar, his adaptedness to a particular line of work, his predisposition to one study or another here receive that thoughtful and careful consideration that are denied the attendants upon larger schools.

Weeks, Cummings, and Company.

Messrs. Weeks, Cummings, and Company, proprietors of extensive steam polishing granite and marble works at No. 51 Main Street in Haverhill and across the Merrimack River in Bradford, invite public attention to the great advantages to the buyer which result from their ample facilities and from their long and extensive experience in the manufacture and sale of monumental work.

They call attention also to the evident fact that the great extent of their business and the convenient location of their steam polishing mill and principal manufactory, between the railway and tide-water, both contribute materially to reduce the cost of manufacturing, handling, and shipping monumental work to the minimum.

They have at all times on hand in their warerooms a large and varied stock of finished monumental work, as well as a complete collection of the most tasteful and practical designs. Correspondence is invited.

The senior member of the firm was the designer of the soldiers' monument, to which reference was made in the earlier pages of this book, which has given general satisfaction to the Haverhill public, and which is a sufficient guarantee of his artistic taste.

Mitchell and Company.

This firm, now consisting of F. J. Mitchell and George Thayer, began business in 1876 with a small stock of goods in a store containing only 1250 feet of

flooring, but has been compelled to increase its space by the demand of a constantly growing business until now it boasts one of the largest and best appointed dry goods houses in Essex County, the making of cloaks being a specialty.

The Merrimack National Bank,

Organized July 5, 1814, can safely claim to be the oldest financial institution in Haverhill. It paid ninety-seven semi-annual dividends, averaging four per cent as a state bank, and as a National bank has averaged semi-annual dividends of five per cent on its capital stock of $240,000. Its officers are: President, C. W. Chase; vice-president, John B. Nichols; cashier, Ubert A. Kil-

lam; directors, C. W. Chase, Moses Nichols, John B. Nichols, Dudley Porter, P. C. Swett, Woodbury Noyes, J. L. Hobson, C. E. Wiggin, John Pilling, C. W. Arnold. The bank's policy has always been the wise one of "regarding wholly the agricultural and manufacturing interests of Haverhill and vicinity in loaning money." Its statement October 4, 1888, showed: Capital stock, $240,000; surplus, $120,000; individual deposits, $410,-000; United States deposits, $105,000. Its deposits averaged, from 1814 to 1850, $6,000; 1850 to 1864, $26,-000; 1864 to 1876, $86,000; 1876 to 1888, $300,000.

Bradford Academy

Is the oldest seminary for young women in the country, founded in 1803, and incorporated in 1804. The school edifice, including the boarding and school department under the same roof, is located near the center of an area of twenty-five acres. The location is high, the air is fresh, sunlight abundant. Pupils have invigorating exercise in the open air, boating and skating on the lake, bowling in the alley, or walking in the grove. The open grounds are laid out in spacious lawns and adorned with shrubs and flowers. Paths are laid through the forest, along the side of the lake, through the dense thickets and the open woods, affording many views of wild and picturesque beauty.

The curriculum includes both classical and English courses of study.

Bradford Academy is in the interest of Christian education. The design is the development of Christian womanhood. By the best nurture, by the choicest instruction, by all personal influence and example, the teachers endeavor to train the pupils for the highest service to which God may call them.

TRUSTEES. — Hon. George Cogswell, M. D., President, Bradford; Ezra Farnsworth, Vice-President, Boston; John Crowell, M. D., Secretary, Haverhill; Doane Cogswell, A. M., Treasurer, Bradford; Rev. John D. Kingsbury, D. D., Bradford; Hon. William A. Russell, Boston; Rev. James H. Means, D. D., Boston; Rev. Edmund K. Alden, D. D., Boston; Elbridge Torrey, Boston; Rev. Nehemiah Boynton, Boston.

CLERK. — Harrison E. Chadwick, A. M., Bradford.

PRINCIPAL. — Miss Annie E. Johnson.

The Second National Bank.

The Second National Bank of Haverhill was chartered May 25, 1886, began business July 1, and in October moved into its present quarters, No. 35 Washington Street, expressly fitted for the purpose. The following were chosen directors: John A. Gale, George A. Greene, Joseph W. Vittum, John Pilling, George H. Carleton, James H. Winchell, George A. Hall, Edgar O. Bullock, John W. Russ, George E.

Elliott, Charles W. Arnold. Mr. John A. Gale, to whose untiring efforts the starting of the bank was mainly due, was elected president; Mr. George H. Carleton, vice-president; Mr. C. H. Goodwin, cashier. Thanks to the efforts of the president, and directors, the bank has pursued a steady, progressive course from the start. Its object has always been and continues to be, to assist in business young men of worth and ability.

211

Saunders Brothers.

Only six years ago, in 1883, Messrs. Albert F. and George Saunders, under the firm name of Saunders Brothers, began business as plumbers and tinsmiths, and dealers in stoves, furnaces, and gas fixtures, starting in a small way, employing but two men. The extent and development of their plant and business may be partially inferred from the fact that they employ eight times as many workmen now. Their salesroom, at No. 9 Emerson Street, fifty feet long and forty feet wide, is handsomely fitted up with all necessary appertenances, is admirably adapted for the exhibition and display of goods in their line, and is, without doubt, one of the finest in the city. The manufacturing is carried on in a two-story building in the rear. They make a specialty of the Highland range and Chilson furnace. Mr. George Saunders retired about a year ago, but the firm name remains the same.

The same attention to every practical detail, the same energy and enterprise, the same honesty and thoroughness in the execution of whatever is entrusted to their hands, that originated and continued the success of the firm, still remain with it and ensure satisfaction.

Haverhill Iron Works.

Some of our enterprising citizens in 1881, realizing the need of a variety of interests to advance the prosperity of Haverhill, formed the corporation known as the Haverhill Iron Works for the manufacture of castings and almost all kinds of finished machinery and heating apparatus. In 1889 the company is doing a business at the rate of $100,000 a year, or more than double what it has ever before done, as our citizens have found out that all their iron work can be done cheaper and better here than elsewhere. It has just dawned upon this community that there is nothing in the line of shafting, machinery, or boiler work, either for power or steam or hot water heating, that cannot be satisfactorily supplied by the Haverhill Iron Works. This company has been looked upon simply as a foundry, and, with the disposition which all citizens have to patronize home industries, no one has ordered columns, store fronts, fire escapes, door steps, hitching posts, gas posts, man-holes, or any other casting any where else, but it never occurred to many of them until now how varied are the capabilities of this institution and that the most intricate machinery that is now being run in our nail factories and shoe shops is made here. No system of heating has yet been devised that equals the hot water plant which this company constructs. It is admirable in every way and gives perfect satisfaction to all who have tried it.

Every kind of piping and repairing is done at the down town office of the company at 82 Washington Street, where the superintendent, Mr. M. S. Holmes, can be found ready to make estimates or contracts for everything in his line.

Perley A. Stone.

Four years ago September 1st Mr. Perley A. Stone commenced business, having previously had an experience of seven years in the employ of Mr. J. H. Durgin. He located at 17 and 19 Railroad Square in the Gardner Block. As business increased he hired additional room on Granite and Washington streets, until January 1, 1887, when he removed to one of the best factories in Haverhill, Sanders' new building, which he now occupies, and in addition the small building adjoining. His specialty is ladies', misses', and children's, men's, boys', and youth's turned slippers. These goods are largely sewed by the " national process ", which, supplemented by his patent method of channeling, makes the strongest seam possible. His business has gained from the first in volume until now he makes as many slippers as any house here. He is fortunate in having associated with him, as a special partner, Mr. Luther S. Johnson of Lynn, who is one of the foremost business men of our sister city and of the country.

214

B. F. Leighton and Co.

In 1878 an enterprise of an entirely new type was inaugurated in Haverhill, when Mr. B. F. Leighton established the first and the only wholesale grocery house in the city. He did at first but a moderate business, about one car-load per week being the usual average necessary to supply the demands of his trade, while now the firm handles weekly four times as much. Two years later, in 1880, the firm name was changed to B. F. Leighton and Company, Mr. Leighton taking into partnership with him Mr. Jackson Webster, a man of energy and experience.

Every article, from the largest to the smallest, from the wooden clothes-pin to the barrel of flour or the hogshead of molasses, every form of merchandise kept by a first-class grocery house, can be found here, and of prime quality. The firm are agents for such houses as Washburn, Martin, and Company, and also for the Silver Spray flour, the best family article milled in the West, which serves to bear out their reputation for honest and reliable goods. They contribute to the satisfaction of the tastes of a large contingent of a grocer's customers by keeping all of the leading brands of tobacco and cigars.

It is a well known fact that they offer every inducement and sell goods at the lowest possible prices. Their trade is far from being confined to Haverhill, but they supply the surrounding country as well. They secured a year ago the services of Mr. Harvey R. Eastman as salesman, a young man well known and liked by the trade, and who the firm are satisfied will do all in his power to make everything pleasant for their customers.

J. H. Winchell and Company.

What an integral part of the life and prosperity of Haverhill the shoe business is, has already been told in this volume, but, perhaps, a clearer idea can be obtained by the ordinary reader from a brief account of one manufactory. The illustrations given are of the shoe manufactory of J. H. Winchell and Company, a firm which makes an average out put of 3700 pairs of shoes, men's, women's, and children's, a day. The Washington Street factory, which is five stories high, covering a lot 125 by 40 feet in area, is devoted to the making of women's and misses' boots and slippers, employs 300 hands, and turns out 2500 pairs of finished shoes each working day.

The Phœnix Row factory, four stories high, 65 by 28 feet in dimensions, turns out men's and boys' buff, calf, and Dongola goods, furnishes employment to one hundred and seventy-five hands, and manufactures 1,200 pairs per diem. In addition, the firm has a factory at Candia, N. H., which has a daily output of

600 pairs. The firm makes a specialty of medium and low grade goods, and its productions are sold in almost every large city in the United States, from Belfast in the East to San Francisco in the West. As may be imagined from the number of hands employed and the vast amount of goods manufactured, the establishment is a great factor in the industrial life of the city, distributing as it does, in the various ramifications incident to so large a plant, nearly $7000 a week in wages. The firm consists of James H. Winchell and Myron L. Whitcomb. Mr. Winchell has been in business in the city, most of the time in the shoe business proper, for thirty-one years and has grown with its growth, prospered with its prosperity. He is a keen business man, energetic and far-sighted, quick to seize an opportunity; and the history of the progress of his business, from a three story building, 60 by 20, employing some seventy-five hands, which he occupied some twelve years ago, to its present enormous proportions, is but an epitome of the history of the city itself. His career is but an exemplification of the possibilities afforded in Haverhill for bright, capable young men to carve out position and prosperity.

Mr. Myron L. Whitcomb, the junior partner, is a young man who has been connected with the firm only two years, but who, by his business ability and shrewdness, promises to become, in the not distant future, one of Haverhill's most prominent and far-sighted business men.

The firm manufactures for the jobbing trade in all parts of the country. The Boston office is at No. 106 1-2 Summer Street.

J. H. LeBosquet and Company.

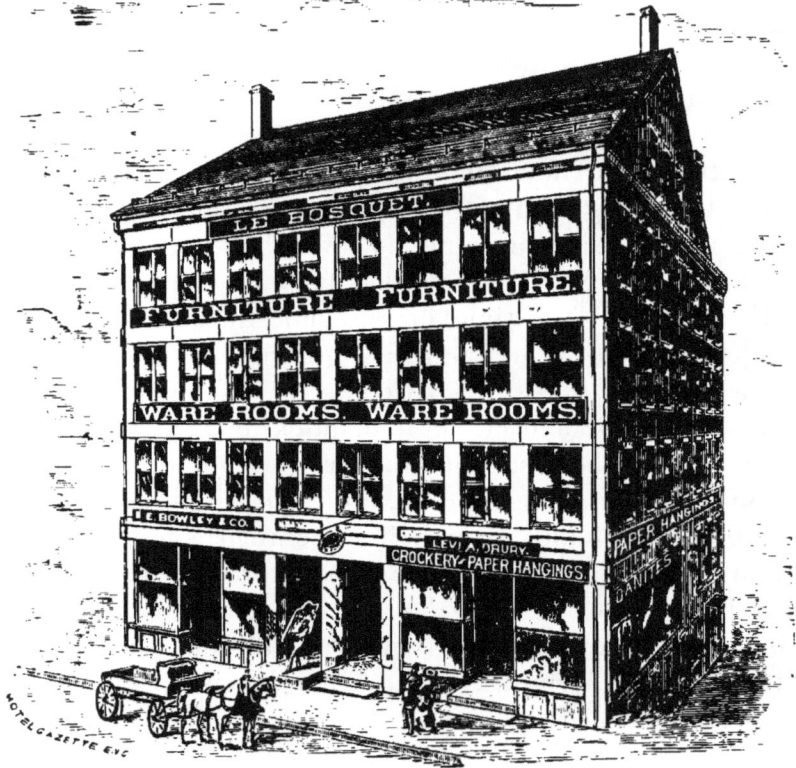

The above is an exterior view of the old and extensive furniture house of J. H. LeBosquet and Company, Nos. 68 to 74 Merrimack Street, affording over 18,000 square feet of floor room. From small beginnings in 1852 the business has steadily increased, until now seven times as much space is required. The same energy, enterprise, and square dealing which gave the firm their start have continued to characterize them since, and their goods are their best advertisement.

H. L. Dole.

H. L. Dole, jeweler, came to Haverhill from Hallowell, Maine, in 1865, and commenced business at No. 4 Merrimack Street, under the firm name of H. L. Dole and Co. Twelve years later the firm ceased to exist, and Mr. Dole became the sole proprietor of the business, which had steadily increased in volume from the first.

In 1879 Mr. Dole removed to his present fine store, occupying the entire first flat at No. 19 Merrimack Street. Mr. Dole has an unexcelled reputation, and his store is frequented by persons looking for first class goods in his line. His establishment is headquarters for all grades of jewelry of the latest and most choice designs. The display is large and complete of watches, gold and plated chains, rings, and solid silver and plated ware of all kinds. Anything that can be found anywhere in a first class jewelry store can be found at this popular establishment. Mr. Dole employs trustworthy clerks, and customers are sure of prompt attention and polite treatment. The optical department, under the management of Mr. E. A. Gage, is a new feature, and spectacles and eye glasses are carefully adjusted so as to give the greatest possible relief to weak or defective vision of all kinds. Particular attention is also given to repairing of watches, clocks, and jewelry, and satisfaction in this line is guaranteed in every case.

Mr. Dole makes all selections and purchases in person, and his large experience enables him to select the best goods, and at the lowest figures.

Brooks Brothers.

The well known firm of Brooks Brothers, now the oldest dry goods house in Haverhill, began business in 1858 at No. 10 Main Street, subsequently moving in 1861 and 1866 as the demands of their increasing business or the tendency of trade suggested. In 1869 they were compelled, to accommodate the growing requirements of the public, to buy a place of their own, leasing the upper stories for other purposes. As time wore on, they needed these for their own use, and now occupy all four stories at No. 20 Merrimack Street, with an annex in the rear, covering an area of nearly ten thousand square feet and yet have none too much room. They carry a stock of dress goods, silks, cloaks, cloakings, domestics, small wares, and carpets not to be excelled this side of Boston.

Besides this immense and varied assortment of goods, which brings the advantages of metropolitan stores within reach of the citizens of Haverhill, and the large space which they have come to utilize for its storage and display, the firm has an abundant force of clerks and all of its dealings with the public are marked by a characteristic spirit of courtesy and fair dealing.

James Busfield.

It is evident to even the most casual observer, that the manufacture of machinery is one that requires a native fondness for mechanical pursuits, a close application to detail, and, when done on a large scale, the command of skilled workmen and extensive facilities. Mr. James Busfield, who succeeded in 1880 the long established and well known firm of E. Everson, doing business in Mechanics' Court and engaged in the manufacture of shoe machinery, rolling mills, strippers, etc., as well as in general repairing of the sort, had the advantage of the plant and the reputation he thus acquired and has carried on a successful business ever since. He has the innate desire of men who are masters of their art to do good work, so that it shall speak well of them.

The increase in the number of his customers and the enlargement of his business have compelled him lately to move his establishment to more commodious and central quarters at No. 66 Phœnix Row.

Mr. Busfield is himself a thorough machinist, has none but first-class workmen in his employ, is able and ready to exercise over them an intelligent supervision, and is therefore able to do his work at the lowest possible figures consistent with good workmanship and a satisfactory job.

In putting up shafting and machinery in the majority of Haverhill factories, Mr. Busfield has come in close contact with our business men; and, from the thoroughness of his work and the strict attention he gives to matters of detail, his business relations with his customers have proved more than satisfactory to both parties.

The Haverhill and Groveland Street Railway

Was built in 1877 from Haverhill to Groveland, three miles, and was equipped with four cars and eight horses, carrying daily about four hundred passengers. Its capital stock was $24,000. It has grown since until now it owns thirty-eight cars and eighty-five horses, with fourteen miles of track, carries daily about twenty-five hundred people, and has a capital stock of $144,000. The immediate management of the road is in the hands of a number of Haverhill's representative business men, as follows: Directors, Hon. Levi Taylor, Ira O. Sawyer, William H. Smiley, Ira A. Abbott, John A. Gale, John A. Colby, Philip C. Swett; president, Ira O. Sawyer; clerk and treasurer, John A. Colby.

The offices of the company are situated in the building shown in the above cut at the foot of Main Street.

Fred G. Richards.

At the age of twenty-one Mr. Richards entered the stable business in partnership with his father, who had bought in 1856 what is now the oldest stable stand in the city, it having been used for that purpose over eighty years. Here can be found anything from a tally-ho coach to a saddle-horse, barges, hacks, carriages, of all sorts, and accommodations for a hundred horses. The facilities for boarding and for transient trade are unusually good. There is a pleasant waiting-room for ladies, a good office, a harness-room, and wash-room connected, all heated by hot water. The stable is never closed, so that an order, by telephone or in person, never fails of attention day or night. Conveniences for hot and cold water, electric bells, electric lights, and telephone combine with the other facilities to make the business the largest in this line and the most complete in this part of the state.

Mr. Richards has not limited his enterprise to the stable business alone, but in 1886 he formed a co-partnership with Mr. G. H. Dole, under the name of Richards and Dole, and bought out the old and well known undertaking establishment of J. H. Cummings. Mr. Dole was brought up an undertaker, serving years at the trade, as his father before him pursued it, so that he united a peculiar fitness for the business with Mr. Richards' extensive livery. The firm have added new equipments and all the modern conveniences, and have obtained a large business and a good reputation not only in Haverhill but in the surrounding country, in which they have many patrons.

The Haverhill National Bank,

Succeeded in 1864 the Haverhill Bank, which was incorporated in 1836. In 1882 it moved into its present elegant rooms in the Masonic Building, fitted up expressly for its use. Besides the greater room needed for its large and increasing business, it has obtained fire and burglar proof vaults, constructed in the very best

manner known to science, and which afford absolute security. The bank's capital is $200,000, surplus fund, $100,000, undivided profits about $25,000. The management means that a liberal spirit of accommodation and a courteous and kindly attention in its dealings with the public shall characterize this bank. The officers are: President, A. Washington Chase; vice-president, John E. Gale; cashier, Benjamin I. Page; directors, A. Washington Chase, Amos W. Downing, Daniel Fitts, John E. Gale, George A. Kimball, John J. Marsh, Eben Mitchell, Thomas S. Ruddock, Thomas Sanders. The uniformly prosperous course of the bank in the past affords reasonable and trustworthy assurance of its continued success in the future.

E. W. Gould.

"All flesh is grass" and all the clothing worn by civilized man becomes in time discolored and soiled. It was the recognition of this fact that first induced Mr. E. W. Gould, proprietor of the Bay State Dyeing and Cleansing Works at 140 Merrimack Street to open his establishment and to ask for the patronage of Haverhill's citizens. Mr. Gould had been in the business for many years in the neighboring city of Lawrence, but recognized the superior advantages offered in Haverhill, where, owing to the fact that in these days, when the shades and colors can scarcely be enumerated, and when the inexorable rule of fashion permits a shade to be popular but one season, the services of a practical dyer are necessary to almost every family, he has obtained a large and constantly increasing patronage, a patronage which has compelled him to add all the modern improvements to his establishment, thereby greatly increasing his facilities for fine work.

It has always been a boast with the establishment, and one reason for its success, that only the best of dyes and chemicals are used and that an experience of thirty-five years of practical work enables it to guarantee its dyeing and cleansing to be equal to that done in the best establishment to be found in the country; and how indeed could it be otherwise, with Mr. Gould with his thirty-five years of experience at the head of the concern, and employing only the best and most careful workmen, under his immediate supervision, in all the ramifications of his business? To have dresses dyed or clothes cleaned by him is to have them renovated, made as good as new.

Hoyt and Taylor.

The firm of Hoyt and Taylor, well and favorably known in Haverhill, consisting of Levi Taylor and Everett Hoyt, began business ten years ago, August 6, 1879. They carry on a very extensive business, both wholesale and retail, in doors, windows, blinds, mouldings, hard wood, fancy lumber, glass, putty, builders' hardware, fancy hardware, sewer pipe, paints, oils, etc.

They occupy the store at No. 152 Merrimack Street, with the building in its rear, besides storehouses, etc., the whole comprising some thirty thousand feet of flooring. They aim to carry in stock everything usually kept in a large and first-class hardware store or required by the needs of carpenters and builders.

The opportunities afforded by this large stock and close attention to business have combined to increase a business at first local by a large out-of-town trade, supplying builders' material from Maine to Connecticut.

The firm has abundant capital at its command as occasion requires, possesses an energetic and sagacious business spirit, and is likely to still farther advance its success. Attentive and courteous in their dealings with customers, its members have obtained the reward that naturally follows.

Ellis and Connor.

The firm of Ellis and Connor, which is composed of Charles A. Ellis and John H. Connor, general partners, and Dudley Porter, special partner succeeded in April, 1887, to the machine sewed business of Goodrich and Porter which latter firm had for years ranked as one of the most substantial and heaviest firms of the city. Their successors are young and enterprising men who seem destined to keep up the high reputation achieved by their predecessors. The specialties of the firm are glazed Dongola button boots in McKay, hand sewed, and Goodyear welt, and they are sold by the case to the jobbing trade of the country from Portland, Maine, on the East, to Portland, Oregon, on the West. The goods manufactured by them have the best reputation for style and quality. Indeed they are Haverhill shoes in the highest sense of the word, which is synonymous with the statement that in all that tends to make perfect footwear they are well nigh unsurpassable.

Their production amounts to four thousand cases, thirty-six pairs in each, per annum, of high grade goods, and this is by no means their limit, as they are steadily pushing onward and their facilities are of the very best, their factory being fitted with all the latest machinery so that all orders are filled with the utmost promptness and dispatch. Their trade mark, E. & C., can be found stamped on shoes for sale in nearly, if not quite, every city in this great country, and when found it is but another advertisement for the city to which this book is devoted, since it is a certain testimony to the skill of its workmen, the judgment and enterprise of its manufacturers, and the reliability and beauty of their products.

W. F. and J. A. Blake.

The business of the above firm was established some ten years ago by Mr. Wilbur F. Blake, who, in 1885, associated with him his brother, J. Albert Blake, under the present firm name. They have several times, by their increasing business, been forced to change to more commodious quarters, and now occupy the entire building shown in the cut, erected by Elijah Fox, and known as the Fox Block. The building itself, one of the most solidly constructed blocks in the city, is, without doubt, the best equipped and best lighted factory in Haverhill.

This firm employs about two hundred of the best skilled operatives in the city, on the higher grades of machine and hand sewed shoes, both in turns and welts. They make the largest number of pairs of fine shoes made by any one factory in Haverhill.

Messrs. Blake control the product of two large factories, one in Calais, Me., known as the St. Croix Shoe Company, under the efficient management of Mr. W. C. Renne, and a factory at Winstead, Conn. Their Boston office is at 22 High Street.

James C. Bates.

One of the best establishments in the city is that of James C. Bates, jeweler, 79 Merrimack Street. Mr. Bates is a native of New Bedford but took up his residence in this city in 1865, where he entered the employ of Kimball and Gould, in which establishment he remained as employee and partner until he entered into business for himself April 27, 1879. For fifteen years Mr. Bates worked at the bench as a watch-maker, and the thorough knowledge of the business thus acquired has stood him in stead since he started business for himself. His establishment contains all the goods that are usually to be found in one of its kind, while the taste and thorough knowledge of the proprietor have been instrumental in building up a large and constantly increasing trade, a trade so large that five workmen are constantly employed in attending to its demands. His success is but another proof of the possibilities which lie before any man in this country who is not afraid to work and who thoroughly acquaints himself with his profession.

Island Stock Farm.

Northern Massachusetts is hardly the ideal place for the establishment of a stock farm. The long severe winters, the variable climate, the herbage itself will hardly compare favorably with that of California or Kentucky. And yet there are stock farms, and good ones, in Massachusetts, farms where some of the best bred and fastest specimens of the trotting horse, at once the pride and enjoyment of the American people, can be found. The little town of Bradford lies on the southern bank of the beautiful Merrimack, just across from Haverhill and it is in this little town that Island Stock Farm, the property of Colonel H. H. Hale, is located. The farm is beautifully situated, the barns and farm house being in close proximity to the river, and in fact derives its name from a large island on the Merrimack used for pasturage purposes. The farm is divided into several sections and contains, in all, some eight hundred acres, under the general direction of Mr. H. L. Burpee, a practical Vermont bred farmer, as superintendent. Island Stock Farm proper contains about two hundred and fifty acres, and on it is situated as fine a collection of stables as can be found on any farm of its kind east of Kentucky, and it is here that the trotting stock is kept, the remaining sections being devoted to cattle, pigs, sheep, and hens, all of the finest breeds and carefully selected. The farm itself is under the highest state of cultivation, and the crops are so extensive, that, despite the enormous outlay incurred by the proprietor, it is practically self-sustaining.

At the head of the stud is Warder, by Belmont. dam Waterwitch by Pilot jr., making him an own brother to Viking. Warder is a bright golden chestnut, five years old and possessed of a world of speed, which he will be, undoubtedly, given a chance to show under the skilful handling of Mr. Byron G. Kimball, the efficient trainer of Col. Hale's stock, and will become standard by performance as well as descent.

Warder, although the premier, is by no means the only stallion at the farm, for dividing the honors with him is Hudson, a four year old bay stallion sired by Kentucky Prince, dam by Rysdyk's Hambletonian. Hudson is a big rangy fellow of much substance and power, and while never trained shows much promise. Bradford Lambert (2.39 1-4), by Addison Lambert, dam Gyp by Redpath, and Comet, by Winthrop Morrill, dam by Champion Morrill, record 2.40 1-4 as a four year old, are the other trotting stallions. There is also an imported Percheron stallion, Major Dome, whose harem consists of three imported Percheron mares, average weight 1700 pounds, and six grade Percherons, the average weight of which is 1500 pounds. There are some twenty brood mares on the farm, including such gilt edged matrons as Silversheen, by Grand Sentinel, dam Peru Belle, an own sister to Strategist, in foal to Warder; Ada Wilkes, by Hambletonian Wilkes, dam the dam of Mambrino Sparkle, in foal to Sultan; Madam Brodhead, by Indianapolis, dam Molly by Magna Charta; Belvidere, by Mambrino Patchen; Kantaka, by Bay State; Lilly Wilkes, by Mambrino Wilkes; Oak Maid, by Almont Eclipse; Wilkesetta, by Young Jim, etc. The list might be prolonged indefinitely, but enough have been mentioned to show that Colonel Hale is breeding only to the best and most fashionable strains.

Thomas H. Bailey.

This pharmacy, located at 23 Merrimack Street, was founded by Mr. George A. Kimball in 1849, and carries on the largest prescription business in the city. The prescription number of this establishment to-day reads upwards of 150,000, which does not include duplicates; had these been numbered, the figures would read 450,000. Over 400,000 of these prescriptions have been prepared since Mr. Bailey became identified with this branch of the business, and he points with pride to this magnificent record.

Floyd and Peabody.

Messrs. Floyd and Peabody are young, energetic men who were brought up in the clothing business, and who are thoroughly acquainted with the demands and needs of the retail trade in their line. The ready made clothing business has, of late years, assumed formidable proportions, and has made vast inroads into the field of patronage formerly held exclusively by the custom tailor. To-day a retail clothier in any large city has to keep in stock goods which, for excellence of material, style of workmanship, fit, and general appearance, cannot be surpassed, at the price, by any first class tailor. There is a large and constantly increasing circle of what is known as the "nobby" trade. But in a city like Haverhill, the metropolis of a large suburban area, there is still another class to be catered to, a class which demand only good articles at reasonable prices and who are not so particular as to style. Moreover, children at the present day are almost invariably clothed by a retail clothier, and the style of their garments is constantly changing, while, to stand the wear and tear to which they are put, only the finest and most substantial cloth can be used. Since their business career commenced, over five years ago, the subjects of this sketch have been indefatigable in catering not only to all these branches of their trade but also to furnishing goods and all the minor details of a patronage which is constantly increasing and which has impelled them to add vastly to their (from the beginning) large and commodious store. They have attained the confidence of the public, and will keep it. Their place of business is at 84 and 86 Merrimack Street.

J. H. Sayward.

The Up Town Hardware Store, of which J. H. Sayward is the proprietor, was established in 1883.

At that time the growth of the city on Mount Washington and vicinity seemed to him to warrant the opening of a store up town, and his judgment proved to be correct. His business has increased to such an extent that where only one man was employed during the first two years, he now employs three besides his book-keeper. His floor surface is far too small to show up the line of goods he carries in as convenient and pleasing a manner as he would like, although he has kept adding to it from time to time as his business would allow, until at the present time he occupies 4300 square feet of salesroom supplied with all the modern conveniences of the present day. His greatest specialty is fishing tackle, and it is conceded by all, that his store is headquarters for anything in that line.

He also carries a full line of builders' and general hardware, farmers' and mechanics' supplies, paints and oils, glass and putty, and everything usually found in a first class hardware store; and he has shown by his push and energy, by a strict attention to business, and by keeping pace with the times, that he merits and has received a generous share of patronage.

The Pentucket Variable Stitch Sewing Machine.

The Pentucket Variable Stitch Sewing Machine is a Haverhill invention, and is owned and controlled almost entirely by Haverhill capital. By its means the possibilities of execution of the sewing machine have been largely increased, since it does easily and completely a class of work which, up to the time of its invention, was done entirely by hand. The machine is a marvel of simplicity, and is constructed according to the most improved methods known in the art of sewing machine manufacture. All the parts are interchangeable and are made of the best material in the best possimanner. No other sewing machine can compete with it in the beauty and excellence of the class of work produced, and an ordinary sewing machine operator can, with a few hours' practice, run it, and can closely imitate all the fancy stitches now made by hand. It will make thousands of fancy stitches without attachments, and a change from one stitch to another can be made instantly while the machine is in motion if desired. It will make a lock-stitch which will not ravel, and silk, linen, cotton thread, or floss of any size can be used. Moreover, the machine works equally well on non-elastic or elastic fabrics, and boots and shoes, corsets, gloves, etc., can be feather or fancy stitched with the greatest ease. As may be seen, the machine fills a long felt want, and its success is not surprising.

Mr. William H. Smiley is the president of the corporation, Charles Howard Poor, secretary, and the home office is in Haverhill.

Hanscom Brothers.

Haverhill is the center of a large agricultural territory, and the firm of Hanscom Brothers has thriven by attention to its wants. The firm, then consisting of M. W. and W. A. Hanscom, bought out in 1865 the long established firm of Paul and Farrington and has since that time been located at No. 30 Main Street, on the

same spot, although a new building has been erected during that time for their occupancy. Their stock comprises paints, oils, hardware, agricultural implements, seeds, etc. Their trade embraces not only a large part of northern Essex but also nearly all of Rockingham County in New Hampshire, and it is no uncommon thing during the spring and summer to see the street in front of their store crowded with the wagons of farmers who have come from ten to thirty miles for the purpose of dealing with Hanscom Brothers. During the first ten years a close attention to business, a keen observation, and a careful consideration of the wants of their trade reaped their natural and legitimate fruit in a five-fold increase of their business, and the growth since has been in the same proportion, steady and constant. The firm now consists of M. W. and J. A. Hanscom.

C. T. Weaver.

Although Haverhill is one of the healthiest cities in that most healthy state, Massachusetts, yet "the wages of sin is death," and no elixir has, as yet, been discovered which will avert inevitable decay and death. Since this is so, and the last sad rites of respect must be paid to our departed ere they return to the dust from whence they sprang, that city is indeed fortunate which can command the services of a competent undertaker and funeral director; and such, there can be no question, Mr. Carlos T. Weaver is. He is thoroughly versed in all the details of his profession, has had years of experience and careful instruction in all its branches, and, moreover, carries a large and complete line of caskets, coffins, robes, etc. His business attained its present proportions only after years of steady industry and personal attention, for it is a business as susceptible of growth as any other, and the qualifications required for success are as great if not greater than in any other.

The confidence of his clients must be gained, confidence in his skill as well as honesty, and this confidence Mr. Weaver has acquired. Haverhill people feel the assurance, that, when his services are required, they will receive just what they want; that the same attention will be paid to the poor as to the rich, and that only in the minor details will the difference be perceptible; and his perfect tact, his sympathy, and attention are at the service of all alike. Both his office and house are connected by telephone, and calls at any hour of the day and night will receive immediate attention. His warerooms are at 34 Main Street.

LeBosquet Brothers.

The business established by C. B. LeBosquet in 1743 has never been out of the family and has come to be regarded as one of the fixtures of Haverhill. It has remained in its present location, No. 20 Main Street, for sixty years. The present building was erected by C. B. LeBosquet, Jr., in 1861. LeBosquet Brothers carry on a general stove, plumbing, furnace, and steam heat-

ing business. They manufacture a low pressure steam heating apparatus, which has been successfully introduced into a large number of stores, public buildings, and private residences, and which has given great satisfaction. In this branch they are wholesalers and retailers, with an office at No. 82 Union Street in Boston. They are agents for the Hub range, and for the Adams and Westlake non-explosive oil stoves, and carry on a large business in gas fixtures and minor articles of trade.

They devote especial care to plumbing, employing only the most expert workmen.

The Academy of Music,

Of which J. F. West is the lessee and manager, and A. A. Ingersoll treasurer, was erected in 1883, and opened to the public on Wednesday evening, September 17, 1885, the opening attraction being the Boston Symphony Orchestra, and an address by Prof. J. W. Churchill of Andover. On the afternoon of July 7, 1888, it was totally destroyed by fire. The work of rebuilding was commenced at once, and in seventeen weeks, on Monday evening, November 12, 1888, it was rededicated by the Redmund and Barry Dramatic Company, who appeared in "Herminie, or the Cross of Gold." In rebuilding, many improvements and alterations were made, making it one of the best appointed theaters in New England. The seating capacity is 1600; proscenium opening, height 41 feet, width 135 feet; depth of stage, 40 feet; width wall to wall, 67 feet; height to fly gallery, 26 feet; width between fly rails, 42 feet; height to rigging loft 55 feet; height of grooves 20 feet. There are three working drops, and thirteen sets of scenery, painted by the well known scenic artist, L. W. Seary, of New York.

The theater is admirably situated, securing eleven exits opening into four streets, the largest audience being able to pass out in three minutes.

Ten months of the year the theater is open, presenting in rapid succession all the leading attractions, consisting of the New York and Boston successes, the leading stars, the great spectacular dramas. Music is not neglected, operas, both the grand and comic, often appearing, the management being always desirous to cater to all the tastes of the amusement loving public.

239

Geo. H. Carleton and Company.

The house of George H. Carleton and Company was established in 1868, under the style of Johnson and Carleton, for the manufacture of ladies' calf and buff shoes. In 1878, Mr. Johnson withdrew from the firm, leaving the business to be continued by Mr. Carleton at the old stand.

In 1880 he removed to his new factory, No. 72 Wingate Street, where he was burned out in the great fire of February 17, 1882. The factory was immediately rebuilt and occupied in July of that year. In 1884 George B. Case became a member of the house, which, under the name of George H. Carleton and Company, has continued the manufacture of ladies' calf unlined and buff shoes for Southern and Western trade to the present time.

This house has always been careful to keep up the quality of its goods, rarely losing a customer, has built up a large and increasing trade, and maintains an excellent reputation as one of Haverhill's representative firms.

John McMillan.

John McMillan came to Haverhill from Boston in March, 1885, and opened an establishment on the upper floor of the Academy of Music. He commenced in an humble way, employing three hands and doing but little business. He paid strict attention to his work, however, and gradually increased his force, until at the present time he gives constant employment to seventeen hands while his business has grown to very large proportions. His first quarters soon grew too narrow, and

he was obliged to move his show and cutting rooms to the lower story, still retaining his former rooms as work-rooms. His present parlors are among the finest in the state, while he carries a full line of cloths such as are usually sold by the best merchant tailors. Mr. McMillan is a good example of his fellow-craftsmen in the city, and his success in establishing so large a business so soon testifies to the character of his work.

The Haverhill Hat Company.

The Haverhill Hat Company was incorporated in 1871, having a paid up capital of fifty thousand dollars, with Eben Mitchell as president and Charles Butters treasurer. At the present time and for the last decade the factory has been running exclusively on orders.

While our predecessors were successful manufacturers, the goods made by them would have but small sales to-day. Some four or five colors and perhaps twenty or thirty styles were all that was then required. Now twenty or more colors and two hundred and fifty different styles are made up for every sale. The Haverhill Hat Company have a wide reputation for their superior colors, acknowledged by dealers to be excelled by no other manufacturer. A specialty during the months of summer and autumn is a line of ladies' and misses' felts. The goods are so well known by the millinery trade throughout the country that the demand is always greater then the supply. In the office of the company hangs the certificate awarded by the International Exhibition at Philadelphia in 1876.

242

Three Taylors.

Above is presented a view of the interior of the clothing house of Three Taylors, 73 and 75 Merrimack Street. The business of this firm, first established nearly a half century ago by the now senior member of the firm, the Hon. Levi Taylor, has constantly grown. From time to time small stores have been given up and larger ones taken to meet the increasing demand for well made clothing, until they now occupy one of the largest stores in Essex County, containing about six thousand feet of floor room. Persons visiting the city should not fail to look through this establishment, where may be found a large assortment of gentlemen's clothing and furnishing goods, suited to the various seasons of our climate and in sizes from the small boys' suit up to that which will fit the extra stout and tall man.

243

Gardner Brothers.

In 1869 Gardner Brothers (E. W. and S. P. Gardner) began the manufacture of ladies' serge shoes in a factory on Washington Square, succeeding the firm of J. Gardner and Sons, which had been in business in Haverhill since 1845. The firm name is unchanged, though Mr. E. W. Gardner has been succeeded by Mr. John H. Thomas, who had been for twenty years superintendent of the factory.

Six years ago the firm built a large and commodious factory, Nos. 38-44 Granite Street, to which the business was removed, and here all of the manufacturing is now done, a part of which, after the fashion of other days, was once done in the country. Gardner Brothers employ about a hundred and fifty hands, and make annually about a quarter of a million pairs of shoes,—men's calf and buff buttons, balmorals, and congress, ladies' kid, Dongola, glove grain, buttons and Polish. They manufacture medium grades, all for the Southern and Southwestern trade, which command a ready and constant sale. The Boston office is at No. 115 Summer Street.

W. B. Thom and Company.

Among the important industries of the city, and second only to the shoe business, is the manufacture of hats, of which the extensive factory of W. B. Thom and Company is the largest. The plant is situated on River Street, a few hundred yards west of the Boston and Maine Railway station, occupying a lot of two hundred and eighty feet front and running back to the river. The plant now comprises five buildings, containing col-

lectively about 32,000 feet of floors, engine-house, boiler-house, store-houses, etc.

The capacity of the original factory was about *seventy dozen* wool hats per day. The present plant, when in full operation, can turn out *four hundred dozen* per day, of every variety of fur and wool hats, for men's, ladies', and children's wear. They employ about five hundred hands when in full operation, with a pay-roll of nearly five thousand dollars per week.

T. S. Ruddock and Son.

The senior member of the firm, Mr. Thomas S. Ruddock, began in 1858 in West Newbury, five miles from Haverhill, the manufacture of men's and women's machine and women's hand sewed slippers. In the spring of 1884 his factory was burned and he came to Haverhill, establishing himself here at 23 and 25 Essex Street. In the fall of that year he associated with himself his son, Mr. Austin E. Ruddock, under the name of T. S. Ruddock and Son.

After coming to Haverhill the business caught the impetus of its surroundings and grew apace, so that in October, 1888, the firm moved again, this time to 130 Washington Street, in order to obtain more room. This factory already, in less than a year, has proved too small, and it is now in contemplation to add another story to accommodate increased demands, although the firm's facilities were greatly enlarged and increased by the change.

Ruddock and Son manufacture men's, women's, and misses' hand and machine sewed shoes and slippers. They are sold exclusively to the jobbing trade, and find a market in all parts of the country, in New England as well as in the South and West.

The long experience of the senior partner in the manufacture of shoes, extending over thirty years, has amply qualified him for the successful management of a large business. His son has grown up with it, and the firm, with the present enlarged facilities and the prospect of more, lacks nothing, apparently, needed for even a more successful continuance. The firm's Boston office is at 112 Summer Street.

Hazen B. Goodrich and Company.

This firm began business in April, 1887, at 72 Washington Street, the senior partner having been for some years a member of the well known and long established firm of Goodrich and Porter. The junior partner, Mr. Frank J. Bradley, was admitted to the firm in July, 1889.

Goodrich and Company manufacture a line of women's hand turned button boots, low-cut shoes, and a Goodyear welt, exclusively for the jobbing trade, and a large and varied line of men's and women's shoes and slippers for the finest retail trade. They also manufacture a patented shoe, which is a hand turned shoe with an extension edge that gives it the appearance of a welt sewed shoe. Their goods include all sorts of Dongola, ooze calf, goat, alligator, plush, embroidered goods, etc.

They occupy a fire-proof factory, spacious, well lighted, fitted with all the conveniences exacted by the modern methods of shoe manufacturing, and their facilities are thus unsurpassed. The results are seen in the products of the factory, which have a reputation for unequalled excellence, completeness, beauty of finish and wearing qualities. They find a ready market in all sections of the country from Maine to Florida, having a large sale in California and Texas. They amply justify the reputation which Haverhill has gained for the manufacture of the best class of goods.

Mr. Goodrich's long experience in the making and selling of this class of goods has given him a peculiar fitness for it, and Mr. Bradley's energy and acquaintance with the trade assist the firm to command success.

First National Bank,

Was organized as the Union Bank in 1849 with a capital of $100,000, increased in 1885 to $150,000, and in 1857 to $200,000. It was re-organized in 1864 as the First National Bank, and in 1870 a stock dividend of 25 per cent was declared and its capital increased to $300,000. Its surplus fund is $120,000, and its undivided profits $20,458. Its aim, to supply the wants of the business men of Haverhill, as demonstrated by the increase of capital as occasion required, is still the policy of the present management. Its officers are: President, George Cogswell; cashier, E. G. Wood; directors, George Cogswell, Levi Taylor, Samuel Laubham, R. Stuart Chase, S. Porter Gardner, Charles C. Griffin, S. H. Gale, James H. Durgin, E. G. Wood.

For many years the bank was located on Merrimack Street, but with the growth of the city westward, a site was purchased on Washington Street and a substantial brick building erected, in which the bank occupies handsomely furnished rooms.

G. W. Pettengill.

Although the character of a place like Haverhill has suffered marked changes in the course of years, and although the village which was once the market place for a wide circuit of surrounding country has now many rivals that divide its commerce, yet some of its characteristics remain unchanged and it is still a natural center for trade and exchange in agricultural products, though many of these at the present time are imported from a distance, instead of being grown, as was the custom formerly, in the immediate vicinity of Haverhill.

Conspicuous among the large dealers in hay, all sorts of grain, and straw, is Mr. G. W. Pettengill, who succeeded in July, 1884, to a business pursued by Mr. E. G. Cheney, and whose place of business is at Nos. 34 and 36 Fleet Street, at what is known as the "old Hunkins stand," where for many years back a lively trade in hay and grain has been carried on. On this spot Mr. Pettengill has remained, continuing the old traditions, doing a large business, averaging over sixty thousand dollars' worth a year, more than three times its volume during the first year, and steadily increasing. His hay he ships from Maine, New Hampshire, and Canada; his grain he brings from the West; his straw from New York state. He uses weekly a car-load of oats and of meal, more than a car-load a week of hay and straw, and a car-load of corn every two or three weeks.

Mr. Pettengill is a Haverhill man born and bred, popular, energetic, ambitious, and deserves the success he has attained.

The Phoenix Drug Store,

Of which Messrs. Frank E. Pollard and Frank E. Watson are proprietors, had its origin in the fall of 1879, when it was rather sneeringly remarked that some insane persons were to open a drug store at the corner of Washington Street and Washington Square, with predictions not very flattering to the young men who were undertaking the enterprise. They were meeting with success, however, when the great fire of the spring of 1882 swept away their store and stock. The store was rebuilt, however, and its present name, "The Phœnix Drug Store," arose from that event.

The retail department is located on the corner of Haverhill's main business thoroughfare, in a large, commodious, and well lighted store. Special attention is paid to the courteous reception of trade.

To the strict attention and personal supervision exercised in the prescription department has been due the marked increase in this branch, which now requires the attention of three experienced pharmacists. The soda and mineral water business has been developed to its present condition by their efforts, they being the first to introduce the Saratoga mineral waters here.

Dermicure, a lotion for the skin, and the Eastern Milk Remedy, known to be successful in the treatment of rheumatism, are manufactured by this firm. They also manufacture fruit juices for soda fountain use by their own peculiar method, which they hold as a secret. These juices, orange in particular, have a reputation that sells them in nearly all parts of the United States. Their laboratory contains the newest machinery and employs the most approved methods.

Hunkins and Wildes.

It is characteristic of Haverhill, and of Haverhill's methods and business, that no man need feel that he must fail in life for lack of an opportunity or on account of his youth. No avenue of success is shut to him for these reasons, and, therefore, the enormous aggregate output of boots and shoes annually sent out from Haverhill is not the product of one or several gigantic factories or large corporations, but represents a total of goods made by several hundred firms, larger and smaller, whose number is every now and then increased by men of experience who have decided to leave the factories of others and strike out for themselves.

Such firms as that of Hunkins and Wildes, though each member had been in business for himself before this partnership was formed, represent this tendency in Haverhill's chief business.

Familiar from youth with the manufacture of shoes, both bring to this comparatively recent association the qualifications for success in long acquaintance with manufacturing processes, and a personal skill in using them. The senior partner, Mr. Warren C. Hunkins, had been for some years a member of the firm of J. B. Swett's Sons, while the junior member, Mr. E. J. Wildes, had been in business alone since 1883.

They formed the present partnership in October, 1888, and established themselves at 25 and 27 Railroad Square. They make a general line of men's, women's, and boys' fine and medium hand sewed shoes and slippers, making a specialty of hand sewed goods and of hand work in distinction from machine work. They are constantly adding new styles and combinations.

Charles Emerson and Sons.

It is not unreasonable to suppose that the character-istics of a city's stores represent, in part at least, the characteristics of its inhabitants, and it is with pride, therefore, that Haverhill's citizens reflect that in Emerson's Bazaar they possess probably the finest store of the sort in New England and that they can find in its stock anything adapted to their varied tastes that could be got in a metropolitan establishment.

The firm deals in china and all sorts of ware, — glass, earthen, silver-plated, in cutlery and kerosene goods, in fancy articles and toys, and housekeeping utensils in general. They are in direct connection with the large potters of the Old World — Haviland, Wedgwood, Minton, Copeland — and there is no sort of ware, American or foreign, that they do not have in stock, or can not furnish at short notice.

They do a very large importing and jobbing business in addition to their extensive retail trade.

Sumner and Chandler.

Among the most enterprising and progressive of the many live firms connected with the shoe and leather trade, this firm occupies a high rank in the business world.

The partners, James S. Sumner and Charles W. Chandler, are men of extended and varied experience in the leather trade and manufacture of bottom stock for boots and shoes.

The firm as at present constituted began business two years ago, since which time their energy and enterprise have borne the fruit of a steady growth in their business, which now takes rank among the most extensive in their line; and their present factory, though double the size of the one occupied one year ago, is crowded to its full capacity, and the firm are already contemplating a still larger increase of facilities in the near future.

The firm's specialty is the manufacture of a full line of bottom stock for boots and shoes, and the product of the factory is meeting with much favor wherever boots and shoes are made. They are the only concern in the city manufacturing fine moulded counters; in this, as in many other things, they show their quick appreciation of the needs of the home as well as foreign trade. The thorough practical training of the members of the firm is shown in every department of their factory. This fact, coupled with skilled employees, improved machinery, and the best of raw material, gives a product which is a credit to the firm and which meets with favor in the trade as shown by their increasing sales.

Fred. W. Peabody.

Mr. Peabody started in business as a music dealer in a small way on Water Street four years ago, but shortly afterwards bought out Mr. Orin W. Tasker's old stand at 208 Merrimack Street, which was the oldest, largest, and best stand in the city, the store having been built by Mr. Tasker expressly for the music business.

Mr. Peabody buys and sells musical instruments, on the instalment plan when desired, exchanges them, and repairs them at short notice. Being a musician and a teacher of music, he is well fitted to select good instruments. He is the exclusive agent for the William Bourne and the A. B. Chase pianos, and also has the largest and best assortment of small instruments in the city.

F. N. Livingston and Company.

It is a characteristic of Haverhill's chief industry, and not its least fortunate one, that it is shared by a large number of active and energetic men, often with but moderate capital, and, also, that it naturally surrounds itself with different forms of manufacturing industry, more or less closely related to the main business.

Among the firms actively engaged in one of these subdivisions of shoemaking is that of F. N. Livingston and Company, a wide-awake, enterprising concern, which manufactures top-lifting, and sole-leather and belting heels, making a specialty of their shanks for ladies' turned boots, and moulded heels. The senior partner is Mr. Frank N. Livingston, who, after sixteen years' experience with the well known firm of Goodrich and Porter, started in business for himself some four years ago, hiring a corner of a small room at three dollars a month, doing all of his own work. The increase of the business has, however, necessitated the enlargement of the firm, the junior partner, Mr. George T. Leighton, having been a member about a year.

The business which four years ago needed but one corner of a room now demands accommodations in marked contrast, and the firm is now located at No. 12 Porter Place, where they keep ten men in constant employment. They dispose of the greater part of the product of their factory out of town, selling largely to customers in New York state and in the distant West. Having met with such marked success as to double their business in the past six months, they mean to double it again in the next six.

J. F. and E. J. Donahue.

The firm of J. F. and E. J. Donahue is one of the firms of young and enterprising business men, neither being as yet thirty years old. The senior member, John F., has been identified with the leather business for the last fifteen years, having been employed by the late Otis W. Butters and other prominent dealers. Edward J., the junior member, has been connected with him about a year, the co-partnership being formed June 1, 1888. Their place of business is at 30 Wingate Street. They manufacture men's and women's out-soles, hand-sewed in-soles, in-soles for Goodyear welts, and all kinds for boots and shoes. They make a specialty of children's out-soles for turned work, counters for turned work and moulding, taps, shanks, etc.

By industry and strict attention to details, this firm have steadily increased their trade, employing a number of hands and doing an extensive business. In fact the increase has been so great that additional room will be required before long.

They take great care in preparing their goods and use only leather of best Union tanneries.

Their machines are all of the latest patterns and best makes, and they spare no expense to produce first class goods.

They fill orders in the shortest possible time and guarantee satisfaction in every instance.

This firm's success in the leather trade affords still another illustration of what youth, when combined with business sagacity, strict industry, and an honorable reputation among business men, can accomplish in Haverhill, even in a comparatively brief space of time.

Nason and Tuck.

Messrs. William Nason and William O. Tuck started in the shoe business in August, 1888, and, although young men now, they have both been identified with the business interests of the city for thirteen years or more, Mr. Nason as a partner in the oldest and largest firm of shoe supplies, and Mr. Tuck as a partner in the largest retail grocery in the city.

They manufacture women's, misses', and children's hand sewed slippers, and get out one of the finest lines for the New England, Western, and Southern jobbing trade, using, in the manufacture of these goods, large quantities of kid, Dongola, goat, ooze calf, and glove calf stock, and give employment to a large number of hands.

Their factory, situated at 49 and 51 Wingate Street, one of the principal streets in the city, is a brick building, four stories high, with the best of light and power. Office and salesroom on the ground floor, also an office 105 Summer Street, Boston.

Although Messrs. Nason and Tuck have been in the shoe business but a year, they have the energy and determination to be one of the leading shoe firms of the city. Their first year's business has been one of satisfaction to themselves, and they trust also to their many customers, as they are men believing whatever they sell, that should they give.

L. C. Wadleigh and Sons.

Among the essentials to the manufacture of boots and shoes are good and well fitting lasts, and these have been supplied to Haverhill for almost half a century by the above firm, which was established by the senior partner in 1841. Mr. Wadleigh began business on Mill Street, at the very opposite end of the city from what is now the business center. He soon removed to Stage Street, however, and afterwards to Mechanics' Court, where he did business many years. When the new Odd Fellows' Building was erected on Main Street, the firm, now L. C. Wadleigh and Son, removed to Washington Street, occupying a building on the site of J. H. Winchell and Company's factory, being one of the pioneers in the movement up town.

In 1879, being in need of larger quarters, they leased the Kimball morocco factory, and took in the junior member of the firm. Burned out in the fire of 1882, they obtained their present quarters on Granite Street, which are entirely inadequate for their business; and, when the lease of this building expires, they will probably erect a suitable structure on the Flanders estate, which the young men have recently bought for the purpose. The firm enjoys a good reputation at home and abroad, having an extensive trade outside of Haverhill. They deal largely, also, in last blocks, of which they have always on hand a large stock, not only in this city, where they have several store houses, but in various parts of the country, stored for seasoning. This is an important part of successful last-making, as good lasts require well seasoned timber.

258

Charles H. Cox.

Mr. C. H. Cox, wholesale and retail dealer in flour, grain, hay, and straw, at 19 Essex Street, and the proprietor of an elevator and mill in Bradford, near Haverhill Bridge, began business in 1880 in a small way. Since then, however, the enterprise has been attended with a steady and vigorous growth. One team and one man were then sufficient for a business that now gives constant employment to six horses and fifteen men.

The elevator and mill, one of the best equipped in the state, is about one hundred feet long, forty feet wide, and three stories high, and has a capacity of about sixty thousand bushels of grain in bulk, and twenty-five hundred tons of sacked grain and flour. It has been refitted by

Mr. Cox with the most approved modern machinery at a cost of about four thousand dollars.

Mr. Cox handled last year about one thousand car-loads of hay, grain, and flour, besides the hundreds of car-loads of meal. His membership of the Boston Board of Trade enables him to buy his grain direct from the West, five to twenty-five car-loads at a time, and thus make the lowest prices.

J. Fred. Adams.

For a city with the extensive business interests that Haverhill has, and its past experience, the matter of fire insurance is an important consideration with its business men. Mr. J. Fred. Adams has been engaged in this business for the past ten years, commencing while with the Haverhill Savings Bank and so continuing until last April, when he retired from that institution and established himself in convenient and comfortable offices in the Daggett Building, Merrimack Street, Rooms 12 and 13. He represents the following standard companies: —

London Assurance Corporation of London, England; Firemen's Insurance Company of Dayton, Ohio; Long Island Insurance Company of Brooklyn, N. Y.; New York Fire Insurance Company of New York; American Insurance Company of Boston; and the Worcester Mutual Fire Insurance Company of Worcester, Massachusetts.

In the life and accident branches of the business he has in the agency the Mutual Benefit Life Insurance Company of Newark, N. J., and the Standard Accident Insurance Company of Detroit, Mich., and aims to give the best satisfaction to his patrons.

His past experience warrants his offering his services to those desiring assistance in making investments, or that feel the need of a practical accountant or auditor.

Western mortgages are largely invested in here, and to those desiring such securities he can offer the 7 per cent guaranteed loans of the Vermont Loan and Trust Company, one of the best of its class.